BASIC CONCEPTS IN POPULATION, QUANTITATIVE, AND EVOLUTIONARY GENETICS

BASIC CONCEPTS IN POPULATION, QUANTITATIVE, AND EVOLUTIONARY GENETICS

JAMES F. CROW
University of Wisconsin — Madison

W. H. Freeman and Company
New York

The diagram on the cover shows the spiraling approach
to equilibrium of three chromosomes involved in
the segregation distortion system in *Drosophila melanogaster*
(see pages 191–193). This manner of graphing the trajectories
is due to Dan Hartl and Brian Charlesworth;
the computer program was written by Bill Engels.

Library of Congress Cataloguing in Publication Data

Crow, James F.
Basic concepts in population, quantitative, and
evolutionary genetics.

Bibliography: p.
Includes index.
1. Population genetics. 2. Quantitative genetics.
3. Evolution. I. Title.
QH455.C76 1986 575 85-15893
ISBN 0-7167-1759-X
ISBN 0-7167-1760-3 (pbk.)

Printed in the United States of America

1 2 3 4 5 6 7 8 9 0 MP 4 3 2 1 0 8 9 8 7 6

To

SEWALL WRIGHT

R. A. FISHER

J. B. S. HALDANE

H. J. MULLER

They have taught me much.

CONTENTS

2 INBREEDING, RANDOM DRIFT, AND ASSORTATIVE MATING 29

3 MIGRATION AND POPULATION STRUCTURE 57

4 SELECTION 70

5 QUANTITATIVE TRAITS 110

6 POPULATIONS WITH OVERLAPPING GENERATIONS 149

7 POPULATION GENETICS AND EVOLUTION 178

APPENDIX
PROBABILITY AND STATISTICS 214

PREFACE

Population, Quantitative, and Evolutionary Genetics grew out of a series of lectures that I have given for several years in the biology core curriculum at the University of Wisconsin — Madison. I have expanded the material but still intend it to be covered in one semester, with time left for additional readings. Though mainly theoretical, the book contains experimental data to provide examples and add realism. The level of treatment is appropriate for advanced undergraduate and beginning graduate students. A knowledge of basic genetics and first-year calculus is assumed. I have supplied an appendix for those students not acquainted with elementary probability theory and statistical methods. Following each chapter, and in the appendix, are questions and problems. I have provided answers at the back of the book.

Current research in population genetics employs advanced mathematical methods that are beyond the reach of most biology students. One of my principal objectives is to present some of these research results in a form that can be understood by those interested in the biology, not the mathematics of population genetics. In many ways this is a shortened, less mathematical, updated version of *An Introduction to Population Genetics* (1970), written with Motoo Kimura. Because I have cited that book so often for more detailed analyses and extensions of concepts presented here, I have abbreviated it C&K.

Chapters 1, 2, and 4 include the standard material of population genetics—random mating, inbreeding, random drift, assortative mating, and selection in diploid, sexually reproducing populations. Whatever novelty there may be is more in the manner of presentation than the content itself. Chapter 3, on population structure and migration, extends the classical Wright island model to include measures more suitable for the application of modern molecular methods. In Chapter 5 I discuss the inheritance of quantitative traits, developing heritability theory from a population genetic viewpoint. Chapter 6 introduces age-structured populations, employing both a discrete and a continuous model. I have given Fisher's concept of reproductive value particular emphasis but go beyond Fisher in showing how to assign an absolute rather than only a relative value to this measure. I hope this chapter will help bridge the gap between genetics and population ecology. In the final chapter I discuss evolution. This subject is so vast that one must be selective; I have chosen particular aspects that I feel are interesting and important.

Chapters 3 and 6 are the most difficult ones in the text. Chapter 6 demands more facility with algebra and calculus than the rest of the book. You can skip these chapters without loss of continuity if you want a brief introduction to the material usually included in a population and quantitative genetics course.

Throughout the book I have emphasized molecular genetics as it bears on population genetics and evolution. I had difficulty deciding how much to say about the theory of transposable elements and multigene families, now being developed at a rapid pace. I decided not to emphasize such work on the grounds that the phenomena themselves are still poorly understood and it is hard to judge which theory will remain most relevant. Therefore I have concentrated on the theories of Mendelian inheritance and molecular genetics that are fully established and of broad generality.

It is my hope that this book will whet your appetite for additional information. Readings in the experimental and observational literature provide an excellent way to round out a course in which this book is the basic text. To this end I have provided an annotated list of references, current textbooks, biographies, and journal articles. I have also included Felsenstein's bibliography, which lists virtually every book and article in theoretical population genetics prior to 1981. For a broader background and an enlarged perspective, you may want to look at some of the classical books and reprints of classical papers.

I am indebted to Michael Simmons of the University of Minnesota and to Bill Engels and Carter Denniston of the University of Wisconsin for numerous useful comments and suggestions. I am also indebted to many students who noticed ambiguities and caught errors. Doubtless many remain, and I would appreciate hearing about them from readers.

BASIC CONCEPTS IN POPULATION, QUANTITATIVE, AND EVOLUTIONARY GENETICS

CHAPTER I

GENES IN POPULATIONS

Classical and molecular genetics are mainly concerned with the nature and transmission of genetic information and how this information is translated into the phenotype. Genetic experiments are designed to be maximally informative for the questions being asked. In contrast, natural populations follow their own rules. It is necessary to devise ways to make predictions and test hypotheses when we have no control over the system. We shall see that with a little ingenuity we can do a great deal of genetic analysis of natural populations despite our inability to carry out controlled experiments.

Population genetics theory is most successful when dealing with simply inherited traits — traits whose transmission follows simple Mendelian rules. Yet many of the most interesting and important traits are not so simply inherited: they depend on several genes, which often interact in complex ways with one another and with the environment. Fortunately the inheritance of most molecular traits, such as restriction fragment length polymorphisms (RFPLs), follows simple Mendelian rules; so with increasing molecular study, simple theory becomes more applicable. In this book we shall first consider simply inherited traits and then, in Chapter 5, how this information can be used to understand more complicated inheritance.

Population genetics includes a large body of mathematical theory, one of the richest and most successful in biology. The useful application of this theory has been greatly enhanced in recent years by an abundance of new molecular techniques. The study of the protein products of single genes and of DNA itself has provided an abundance of data to which mathematical theory is immediately applicable.

The theory of population genetics uses advanced mathematics and has become very sophisticated. The subject has attracted many excellent mathematicians, and the literature is now beyond the reach of most biologists. In this book I cannot hope to do more than skim the surface of these mathematical techniques. Fortunately we can get a good grasp of population genetics without advanced methods.

1-1 MEASUREMENT OF ALLELE FREQUENCY

The Gene Frequency Concept. Usually the first step in the study of a natural population is to determine the frequency of different genotypes or phenotypes in the population or, more often, in a sample of the population. From this sample the allele frequencies are determined. Usually we are more interested in the frequency of different alleles than in the frequency of different genotypes or phenotypes. Why?

One reason is that allele frequency is a more economical way to characterize the population. The number of possible genotypes from only a few loci is enormous. Suppose there are 100 loci, each with four segregating alleles. With four alleles 10 different genotypes are possible at each locus: four homozygous and six heterozygous. When all 100 loci are considered together, there are 10^{100} possible genotypes, more than the number of elementary particles in the Einsteinian universe. Yet we can satisfactorily characterize the population for most purposes by specifying only 400 allele frequencies.

The characterization of a population in terms of allele frequencies rather than genotype frequencies has another advantage. In a Mendelian population the genotypes are scrambled every generation by segregation and recombination. New combinations are put together only to be taken apart in later generations, with the result that each genotype is essentially unique, perhaps never to recur in the whole species history. If we are to consider a population over a time period of more than a few generations, the stable entity is the gene (or perhaps a closely linked cluster of genes or a group of genes locked together by an inversion). From this genetocentric viewpoint the basic evolutionary process is the change of gene frequencies.

Allele frequencies may change by mutation, migration, random fluctuation, and selection. Biologists and philosophers have discussed repeatedly and vociferously what *the* unit of natural selection is in a sexually reproducing population. Clearly the organism is what survives and reproduces (or fails to). Natural selection acts on organisms, whose fitnesses determine their likelihood of survival and reproduction. But what the individual transmits to future generations is a random sample of its genes, and those genes that increase fitness are disproportionately represented in future generations. There is selection *for* organismic properties, such as vigor and fertility, and selection *of* genes determining these properties. Whether the unit of selection is a few nucleotides, a gene, a linked cluster, a genotype, a phenotype, or a group of individuals depends on the questions we are asking and the time scale over which we are interested.

A Note on Terminology. The words *gene, locus,* and *allele* are not used consistently in the genetic literature. I shall use **locus** to designate a chromosomal location and **allele** to designate one of the alternative states of the gene occupying this position. However, the use of *gene* as a synonym for *allele* has ample precedent and I shall frequently interchange these words when there is no danger of ambiguity. Therefore gene frequency and allele frequency are synonymous. Furthermore molecular advances have complicated these definitions. Do we regard leader and trailer regions as part of the gene or should we include only the translated part? Are intervening sequences part of the gene? How do we characterize base changes in highly repetitive DNA? The solution I shall adopt is to use the familiar word *gene* when we do not need to be specific and to use more specialized words when we require a finer distinction.

The MN Blood Groups as an Example. Because the blood groups are easily investigated and simply inherited, they have long been a favorite object of anthropological studies. In addition, abundant worldwide data from human and other mammalian populations are available. Table 1-1 shows an example of blood group frequencies in a population.

At a glance we can clearly see that fewer M than N alleles are in this sample. We can verify this observation by direct gene counting. Humans, being diploid, have two

TABLE 1-1

Frequencies of MN Blood groups in a sample from the New York City black population.

BLOOD GROUP	M	MN	N	TOTAL
Genotype	*MM*	*MN*	*NN*	
Number	119	242	139	500
Proportion	0.238	0.484	0.278	1.000
	$P_M = 0.480$	$P_N = 0.520$		

Data from Mourant et al. 1976.

representatives of each gene in each cell. In this sample 119 persons have two *M* alleles and 242 have one. Adding the number of *M* alleles, we get 2(119) + 242 = 480. Likewise, adding the number of *N* alleles, we get 520. The total number of *M* and *N* alleles is 480 + 520 = 1000. The proportion of *M* alleles, which we designate p_M, is 480/1000, or 0.48. The proportion of *N* alleles, p_N, is 0.52. Note that the sum of the allele proportions is 1, as expected.

We are almost always more interested in the proportions of the different alleles than in their absolute numbers. Ordinarily the word *frequency* means either a number or a proportion, but in population genetics we almost always use it as a proportion. I shall adhere to this time-honored custom and use frequency as a synonym for *relative* frequency, or proportion.

Notice that we could have as easily obtained the allele frequency from the proportions of the three blood groups. To get p_M, we simply add the proportion of *MM* to half the proportion of *MN*. Thus $p_M = 0.238 + 0.484/2 = 0.48$, as we obtained previously.

The data in Table 1-1 represent a sample from a population. We regard this sample as representative of the entire population from which it came. Sampling involves many practical problems: How representative is the sample? Was it obtained by a randomizing process? Is the sample size adequate? How large are the standard errors? In this book we shall largely ignore such details, important as they are, so that we can concentrate on general principles.

Table 1-2 gives some additional data on human blood groups. Note that the frequencies vary widely in different parts of the world. In most populations the two alleles occur with roughly equal frequency, but *M* is more common in Central and South American Indians whereas *N* is more common in the regions around Australia. You might enjoy glancing through the book by Mourant et al. (1976), fully referenced at the end of the text. It includes extensive tables of blood group frequencies, serum proteins, and hemoglobins throughout the world.

TABLE 1-2

Additional data on frequencies of MN blood groups in human populations.

	NUMBER					
	MM	MN	NN	TOTAL	p_M	p_N
New Guinea highlands	2	32	269	303	0.059	0.941
New South Wales	3	44	55	102	0.245	0.755
New York City whites	287	481	186	954	0.553	0.447
Guatemalan Indians	112	74	17	203	0.734	0.266
British Columbian Indians	89	31	2	122	0.857	0.143

Data from Mourant et al. 1976.

More Than Two Alleles. The principles of allele frequency discussed in the last section can be readily extended to accommodate any number of alleles. For example, the enzyme locus esterase-2 usually has several different alleles coexisting in the population. Although the protein products of these alleles are indistinguishable in their enzymatic properties, different amino acids can change the electric charge or alter the molecular shape. The molecules then migrate at different rates in an electric field and can be distinguished by gel electrophoresis.

Table 1-3 shows an example of multiple alleles from a population of field mice, *Peromyscus polionotus*, trapped in Georgia. The three alleles are designated by the letters a, b, and c. The proportions are computed exactly as before: for example, $p_b = [2(53) + 13 + 21]/2(97)$.

Because genotypes are customarily given as proportions rather than as numbers, the following rule is usually used to determine gene frequencies:

> **The frequency of an allele is obtained by adding the frequency of the homozygote for the allele to half the frequencies of all heterozygotes involving the allele.**

Notice a very simple consequence of our definition of allele frequency: it must always lie in the range of 0 to 1. If an allele is absent, its frequency is 0; if it is the only allele, its frequency is 1. The allele frequencies must add up to 1.

As mentioned at the beginning of this section, we can greatly simplify the description of a population by using allele frequencies rather than genotype frequencies. In exchange for this simplification we lose information. If we know the frequencies of all of the different genotypes, we can compute the allele frequencies, but we cannot do the reverse. We need additional information on how the genes are put together to produce gametes and zygotes—information about the system by which mates are chosen, linkage relations among the genes, differential mortality, and so forth.

Equations dealing only with allele frequencies are incomplete unless we provide additional information or make additional assumptions. One assumption that greatly simplifies the analysis and usually provides an excellent approximation to reality is discussed in Section 1-2.

TABLE 1-3

Frequencies of esterase-2 genotypes in field mice *(Peromyscus polionotus)*. The three different alleles detected by electrophoresis are *a*, *b*, and *c*.

GENOTYPE	*aa*	*bb*	*cc*	*ab*	*ac*	*bc*	TOTAL
Number	2	53	5	13	3	21	97
		$p_a = 0.103$		$p_b = 0.722$		$p_c = 0.175$	

Data from R. Selander et al. 1971. University of Texas Publication 7103:54.

1-2
RANDOMLY MATING POPULATIONS

A Digression into Model Building. Any description of nature—verbal, pictorial, or mathematical—is necessarily incomplete. Sometimes the description is only a simple caricature or model, but such models can make the description more vivid and understandable. We find it convenient to picture gas molecules as elastic spheres, atoms as miniature planetary systems, and DNA molecules as twisted ladders. We can also use such models to help visualize concepts in population genetics.

One of the most useful conventions in population genetics is the model of a **gene pool.** We assume that each parent contributes equally to a large (theoretically infinite) pool of gametes. We regard each offspring as a random sample of one egg and one sperm from this pool and assume that the parent and offspring generations are distinct. Each gene is chosen randomly, as if we were drawing different-colored beans from a bag. This kind of model building is called—derisively by some, affectionately by others—"beanbag genetics." You might enjoy reading J. B. S. Haldane's amusing and spirited article, "A Defense of Beanbag Genetics" (1964).

Clearly the beanbag model is a crude representation of nature. A real population has individuals of all ages: some dying, some choosing mates, some giving birth, and some simply growing older. Keeping track of all this information is not only impractical, but is usually not very interesting or important. Most questions of genetic or evolutionary interest do not require such minute details. The gene pool model is usually as accurate as the data deserve and much easier to comprehend. It is also much more easily managed algebraically. Using the model we shall find it easy to derive formulae for the relation between gene and genotype frequencies, for the effect of chance in small populations (gene frequency drift), and for selection. The gene pool model will give us the same kind of insights that simple models in the physical sciences do, and often with predictions that are quite accurate—in any case sufficiently accurate for most uses. Departures from the simple model can usually be treated with appropriate modifications. For example, in Chapter 4 we consider the effects of selection by having the parents contribute unequally to the gene pool, that is, in proportion to their "fitness." We treat inbreeding by introducing correlations between uniting gametes.

As I have already emphasized, the simple models we shall discuss are only approximations. They are like the complete vacuum, frictionless pulleys, and perfectly elastic particles that are assumed in physics. We expect to have to modify formulae derived from these models in many practical situations, such as for predictions in animal and plant breeding. The corrections are often empirical and usually do not contribute much to our understanding of the basic ideas of population genetics.

The most unrealistic feature of the gene pool model is the assumption of discrete, nonoverlapping generations. This applies to annual plants and insects but not to most other species. Nevertheless the algebraic simplicity and the qualitative understanding the discrete-generation model provides make it particularly desirable. Much of the newer research in population genetics incorporates more realistic models. We shall consider one such model in Chapter 6.

The Hardy-Weinberg Principle. As an example of the gene pool model, look again at the data on MN blood groups in Table 1-1. In this population the proportion of *M* alleles, p_M, is 0.48 and p_N is 0.52. We assume that a pool of genes exists in these proportions and we draw pairs, corresponding to one gene in an egg and one in a sperm, at random. The probabilities of drawing the four gene combinations are as follows: $(0.48)^2$ for two *M* genes, (0.48)(0.52) for *M* in the egg and *N* in the sperm, (0.52)(0.48) for *N* in the egg and *M* in the sperm, and $(0.52)^2$ for two *N* genes. Therefore we would expect 0.230 *MM*, 0.499 *MN*, and 0.270 *NN* genotypes.

The numerical example illustrates the Hardy-Weinberg principle. The principle says that if the *M* and *N* alleles are in the proportions p_M and p_N, then after one generation of random mating the three genotypes, *MM*, *MN*, and *NN*, are in the proportions p_M^2, $2p_Mp_N$, and p_N^2, as illustrated in Figure 1-1.

Now compare the numbers expected from the Hardy-Weinberg principle with the observed numbers. The total number of individuals is 500, p_M is 0.48, and p_N is 0.52; so the expected number of *MM* genotypes is $500p_M^2$, or 115.2. The observed and expected numbers are given in Table 1-4.

The agreement between observation and hypothesis is very good in this example; the probability of obtaining random results that deviate from the expected results by as much as or more than do the observed results is just about 0.5. If we repeated this

FIGURE 1-1. The results of random combinations of gametes from a gametic pool.

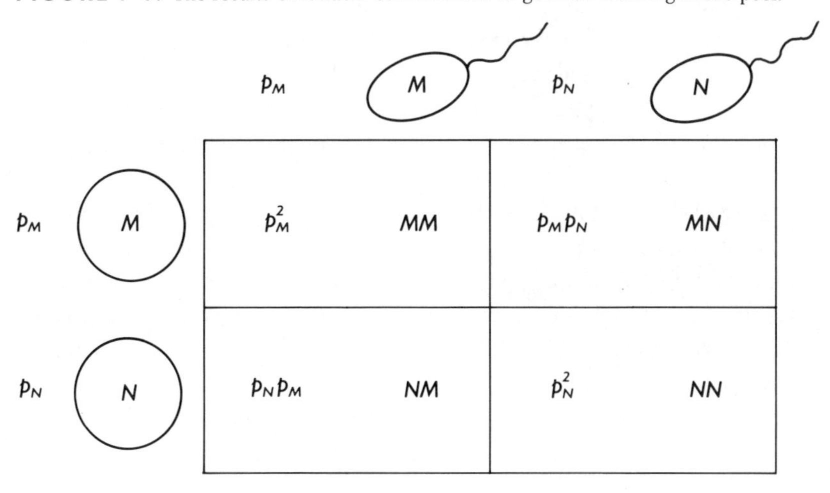

TABLE 1-4

Agreement of the observed number of MN blood groups with the Hardy-Weinberg expectations.

BLOOD GROUP	MM	MN	NN	TOTAL
Observed number	119	242	139	500
Expected number	115.2	249.6	135.2	500

$\chi^2 = 0.46$ $df = 1$ $P = 0.5$

Note that the chi-square method for testing agreement between expectation and observation is described in Appendix 3. You might have expected the number of degrees of freedom to be two, since there are three categories. The number has been reduced by 1 because one independent quantity, the frequency of the *M* allele, has been estimated from the data. Although both p_M and p_N were estimated, only one of them is independent, since if you know one you can compute the other by subtracting from 1.

Data from Mourant et al. 1976.

"experiment" many times, we should expect as bad or worse an agreement between observation and expectation half of the time. You may want to compute the expected numbers in the other populations in Table 1-2 and verify that all are in substantial agreement with the expected proportions.

You are probably impressed, and possibly surprised, by the close agreement between the observed data and the expectation. The simple model of drawing random gametes from a gene pool accurately predicted the genotype frequencies in an actual population. Why did the model work so well? The actual matings in the population must closely resemble a random combination of gametes as far as the *MN* locus is concerned. We are not too surprised at the similarity when we realize that most people are born, grow up, choose a mate, reproduce, and die without ever knowing their MN blood type. The frequencies of the different combinations of matings involving M and N blood groups are determined by the frequencies of these groups in the populations.

Random mating does not mean promiscuous mating. It means only that mates are chosen without regard to genotype at the locus under consideration. A population can be in Hardy-Weinberg ratios at one locus and not at another if mates are chosen according to their phenotype at the second locus but not at the first.

The example in Table 1-4 illustrates the robustness of the Hardy-Weinberg ratios. The black population of New York City is highly heterogeneous. Many are recent migrants from different parts of the United States and the rest of the world. There are variable amounts of white admixture, and the population is unlikely to be mating strictly at random. Yet as far as M and N blood groups are concerned, the population exhibits Hardy-Weinberg ratios.

Although it may seem unnecessary, we can easily demonstrate that random choice from a gamete pool is equivalent to random mating among the various genotypes. Let

TABLE 1-5

Demonstration that random mating among genotypes leads to Hardy-Weinberg ratios. P_{11}, $2P_{12}$, and P_{22} are the proportions of genotypes A_1A_1, A_1A_2, and A_2A_2.

MATING	FREQUENCY OF THE MATING	PROGENY		
		A_1A_1	A_1A_2	A_2A_2
$A_1A_1 \times A_1A_1$	P_{11}^2	P_{11}^2		
$A_1A_1 \times A_1A_2$	$2(P_{11})(2P_{12})$	$2P_{11}P_{12}$	$2P_{11}P_{12}$	
$A_1A_1 \times A_2A_2$	$2(P_{11})(P_{22})$		$2P_{11}P_{22}$	
$A_1A_2 \times A_1A_2$	$(2P_{12})^2$	P_{12}^2	$2P_{12}^2$	P_{12}^2
$A_1A_2 \times A_2A_2$	$2(2P_{12})(P_{22})$		$2P_{12}P_{22}$	$2P_{12}P_{22}$
$A_2A_2 \times A_2A_2$	P_{22}^2			P_{22}^2
Total	$(P_{11} + 2P_{12} + P_{22})^2 = 1$	$(P_{11} + P_{12})^2 = p_1^2$	$2(P_{11} + P_{12})(P_{12} + P_{22}) = 2p_1p_2$	$(P_{12} + P_{22})^2 = p_2^2$

P_{11}, $2P_{12}$, and P_{22} be the parental frequencies of genotypes A_1A_1, A_1A_2, and A_2A_2, respectively, and consider all possible matings. (Designating the frequency of the heterozygous class by $2P_{12}$ instead of P_{12} is just a convenient device for eliminating troublesome 1/2's and 1/4's as coefficients, thereby making the algebra a little easier.)

The calculations are set forth in Table 1-5. Notice that by using the gene-counting method of the previous section (for example, $p_1 = P_{11} + P_{12}$), we find that the progeny are in Hardy-Weinberg ratios. We can state the Hardy-Weinberg principle in symbols as follows:

$$P_{11} = p_1^2 \qquad 2P_{12} = 2p_1p_2 \qquad P_{22} = p_2^2. \tag{1-1}$$

It is convenient to use capital letters for genotype frequencies, lowercase letters for allele frequencies, and subscripts for alleles. Remember that P_{12} is half the frequency of the A_1A_2 heterozygotes.

Since we did not assume that the parental genotypes were in random-mating proportions, this derivation shows that Hardy-Weinberg ratios are attained in a single generation of random mating. This attainment of equilibrium in a single generation is in striking contrast to most equilibria in science, which are approached gradually. This property is what makes the Hardy-Weinberg principle so useful in population genetics. Regardless of the degree of departure from random proportions in the parent generation or how long this departure has existed, one generation of random mating produces

Hardy-Weinberg ratios. In this respect the population has no memory. No wonder Hardy-Weinberg ratios are so regularly found in nature. Note, however, that the allele frequencies in the parent generation must be the same in the two sexes.

If we expand $(p_A + p_a)^2$, we obtain $p_A^2 + 2p_A p_a + p_a^2$. Therefore another way of stating the Hardy-Weinberg principle is to say that the frequencies of the zygotic genotypes correspond to the terms in the binomial expansion of the allele frequencies.

The general principle for randomly mating populations was discovered independently by G. H. Hardy and Wilhelm Weinberg in 1908, and by several others at various times. The principle is a simple application of the binomial theorem; so almost anyone who considered the problem might have discovered it. Hardy was one of the world's greatest mathematicians; you might enjoy his charming, but idiosyncratic, essay "A Mathematician's Apology" (1940). Weinberg solved many problems of much greater difficulty and devised a number of clever tricks for studying human genetics. Paradoxically both of these men are best known for their least profound discoveries.

Despite its simplicity, the Hardy-Weinberg principle forms the basis for almost all analyses of natural diploid populations. The single-generation approach to equilibrium and the approximate random mating that characterizes most populations ensure its applicability. We can simplify the study of natural populations by using allele frequencies, knowing that whenever we wish we can convert them into zygotic frequencies by the Hardy-Weinberg principle.

Some Additional Examples. If alleles exhibit dominance, we cannot recognize all the genotypes from their phenotypes. A good example is the severe human disease cystic fibrosis. It is caused by homozygosity for a recessive allele, and the proportion of affected newborns is about 1/2500. Let a and A stand for the abnormal and normal alleles, and p_a and p_A for their frequencies. Then with random mating, $p_a^2 = 1/2500$. Therefore $p_a = 1/50$ and $p_A = 49/50$. The expected proportion of *Aa* heterozygotes, who are normal but carry the defective allele, is $2p_A p_a$. Numerically this proportion is 98/2500, or about 1 in 25. In a class of 100 genetics students, 4 would be expected to be carriers of this allele.

But, is the random-mating assumption reasonable in this case? In one sense it clearly is not, for homozygotes hardly ever reproduce. Nevertheless, the *AA* and *Aa* genotypes mate at random; so if we consider the gene pool from reproducing genotypes, the random-mating model is appropriate. The allele frequencies of newborns are the same as those of reproducing adults in the previous generation (if we ignore mutation). This raises another question. Since the *aa* genotype makes no contribution to the gamete pool, we would expect a slow decrease in *a* alleles unless this reduction is balanced by some other process, such as mutation. This question involves selection, which we shall take up in Chapter 4.

The MN blood groups do not noticeably affect health or survival. Hence we would expect Hardy-Weinberg ratios in persons of all ages (barring such complications as

TABLE 1-6

Observed and expected frequencies of chromosome inversion types in *Drosophila pseudoobscura*.

	NUMBER				CHROMOSOME FREQUENCY			
	ST/ST	*ST/CH*	*CH/CH*	TOTAL	*ST*	*CH*	χ_1^2	*P*
Eggs observed	41	82	27	150	0.547	0.453		
Eggs expected	44.8	74.4	30.8	150			1.57	0.2
Adults observed	57	169	29	255	0.555	0.445		
Adults expected	78.5	126.0	50.5	255			29.7	<0.0001

Data from Th. Dobzhansky. 1947. *Genetics* 32:146.

young immigrants coming from a population with different allele frequencies). With cystic fibrosis, the *aa* class is deficient in older groups because of its lethality.

The data in Table 1-6 illustrate the effect of differential mortality on Hardy-Weinberg ratios. In this case we are dealing with an entire chromosome instead of a single locus. The two chromosome types, characterized by different inversions, are called standard *(ST)* and Chiracahua *(CH)* and were originally derived from wild populations of *Drosophila pseudoobscura*. The three chromosomal genotypes are easily identified by analysis of larval salivary gland chromosomes.

The egg data were obtained by collecting fertilized eggs from a caged population and then letting them develop into larvae under very favorable conditions, with abundant food and no competition. Note that in the table the larval genotypes (and therefore the eggs from which they developed) are very close to Hardy-Weinberg expectations. On the other hand, the adults were sampled directly from a crowded cage, and their genotypes were determined by progeny tests (salivary gland chromosomes are visible only in the larval stages). The conclusion is clear: the intense competition of the cage caused high mortality, resulting in a very large departure from Hardy-Weinberg ratios. Notice that the chromosome frequencies hardly changed at all between the eggs and the adults. The departure from expectation in the adults is caused by a deficiency of both homozygotes. The excess mortality is not caused by the chromosome inversion per se but by homozygosity for deleterious recessive genes that are "locked in" by the crossover-suppressing effects of the inversions.

This example illustrates another aspect of the Hardy-Weinberg principle. It need not be confined to a single locus. It applies to a group of linked genes, to a chromosome, and even to an entire gamete. In generalized form the principle states that the zygotic frequencies of whatever is being considered are given by the binomial expansion of the gametic frequencies.

Multiple Alleles. The extension of the Hardy-Weinberg principle to multiple alleles is straightforward. Table 1-7 shows the same data as were given in Table 1-3. The allele frequencies, as previously obtained, are $p_a = 0.103$, $p_b = 0.722$, and $p_c = 0.175$. By analogy with the two-allele case, we would expect the proportion of *aa* homozygotes to be p_a^2, or 0.0106, with the other two homozygotes calculated in the same manner. The expected *ab* heterozygote frequency is $2p_ap_b$, or 0.1487, with the other heterozygote frequencies calculated similarly. The expected numbers can be determined by multiplying these proportions by the total number of mice, 97, and are given in Table 1-7. The agreement between observation and expectation is good, showing that the three-allele extension of the Hardy-Weinberg principle accurately accounts for these genotype frequencies. Since the data are based on adult mice, we can conclude not only that mating is at random with respect to this esterase locus but also that no measurable differential mortality is associated with these genotypes.

We can summarize these calculations and generalize them to accommodate any number of alleles by the following rules:

1. **The frequency of any homozygous genotype is the square of the corresponding allele frequency.**
2. **The frequency of any heterozygous genotype is twice the product of the two component allele frequencies.**

TABLE 1-7

Observed and expected frequencies of esterase genotypes in a population of field mice (data from Table 1-3).

	GENOTYPE							ALLELE FREQUENCY		
	aa	*bb*	*cc*	*ab*	*ac*	*bc*	TOTAL	p_a	p_b	p_c
Observed	2	53	5	13	3	21	97	0.103	0.722	0.175
Expected	1.0	50.5	3.0	14.4	3.5	24.5	96.9			

$\chi_3^2 = 3.0 \qquad P = 0.39$

Note that the number of degrees of freedom is $5 - 2 = 3$, since two independent allele frequencies were determined from the data. Once two allele frequencies have been determined, the third is fixed by the necessity that they total 1. The chi-square test violates the practical rule that expected numbers should not be less than 5. (Actually the rule may be unnecessarily conservative; using 3 instead of 5 usually gives a satisfactory approximation. See Cochran, 1954.) If the *aa* and *cc* classes are combined, the observed number is 7 and the expected number is 4.0. Using these values, we get $\chi_2^2 = 2.9$, $P = 0.23$. Notice that as we reduce the number of classes, we reduce the number of degrees of freedom correspondingly.

We can express the same relationships between genotype and allele symbolically, which will also introduce the symbols used throughout the book. We designate the n alleles by $A_1, A_2, \ldots, A_i, \ldots, A_j, \ldots, A_n$, and their frequencies by $p_1, p_2, \ldots, p_i, \ldots, p_j, \ldots, p_n$. In this notation, the subscript i designates any allele and j any other allele. As before, using capital letters to designate genotype frequencies and lowercase letters for alleles, we have

$$P_{ii} = p_i^2 \qquad P_{ij} = p_i p_j \qquad H_{ij} = 2P_{ij}. \tag{1-2}$$

For convenience we let H_{ij} stand for the frequency of heterozygotes involving alleles A_i and A_j, and P_{ij} for half this value.

We are often interested in what proportion of the population is homozygous or heterozygous at a particular locus. We can get this information by simply adding the frequencies of the component genotypes, as shown in the following equations:

$$\text{Homozygotes:} \qquad p_1^2 + p_2^2 + \cdots + p_n^2 = \sum_{i=1}^{n} p_i^2;$$
$$\text{Heterozygotes:} \qquad p_1p_2 + p_2p_1 + p_1p_3 + \cdots = \sum_{i \neq j} p_ip_j = 1 - \sum_i p_i^2. \tag{1-3}$$

Of course we can more easily calculate the proportion of heterozygotes by computing the proportion of homozygotes and subtracting from 1. There are $n(n-1)/2$ heterozygotes, but only n homozygotes.

Multiple Alleles with Dominance. Given the allele frequencies and assuming random mating, we can compute the genotype frequencies. Calculating the allele frequencies from the genotype frequencies can be much more complicated if dominance causes the same phenotype to be produced by more than one genotype.

Table 1-8 gives some data on the frequencies of different-colored snails collected from a natural population in Britain. The three alleles, in order of dominance, are B, b', and b. We can estimate the allele frequencies as follows: the frequency, r, of the bottom recessive is obviously the square root of the proportion of yellow snails, 0.197, or 0.444. We then note that the sum of the yellow and pink snails, 0.587, is $(q + r)^2$, but $q + r = 1 - p$. Therefore p is $1 - \sqrt{0.587}$, or 0.234. Knowing p and r, we can determine q by subtracting $p + r$ from 1.

Although in this case we easily found a way to estimate the allele frequencies, it is usually more difficult. The standard method used by population geneticists for more complex problems is the method of maximum likelihood, which is explained in Appendix 4 using this same example.

TABLE 1-8

Frequencies of three color types in a natural population of snails, *Cepaea nemoralis*, in Britain. The frequencies of alleles B, b', and b are p, q, and r, respectively.

COLOR	NUMBER	PROPORTION	GENOTYPE	EXPECTED PROPORTION
Brown	88	0.413	BB, Bb', Bb	$p^2 + 2pq + 2pr$
Pink	83	0.390	$b'b'$, $b'b$	$q^2 + 2qr$
Yellow	42	0.197	bb	r^2
Total	213	1.000		1
		$p = 0.234$ $q = 0.322$	$r = 0.444$	

Data from A. Cain et al. 1960. *Genetics* 45:406.

Frequency of Heterozygous Loci in Natural Populations. Until recently we could not estimate the total amount of heterozygosity in natural populations. For loci such as the blood groups, that number has long been known, but for the thousands of other loci, almost no information had been available. Certainly a great many segregating loci for quantitative traits must exist in any population. How else could we account for the enormous variety of sizes, shapes, colors, and so forth that characterize all natural populations? But whether or not the large absolute number of segregating loci represents a large fraction of all loci has been totally unknown until recently.

Before the development of molecular genetics, speculation about the total amount of heterozygosity ran the gamut from those who asserted that the great majority of loci, say 95 percent or more, in any particular organism were homozygous to those who asserted, just as confidently, that 95 percent or more were heterozygous. For those loci producing easily studied enzymes, considerable information is now available, and some examples are given in Table 1-9. These data suggest that those arguing for a minority of heterozygous loci were more nearly correct.

We must distinguish between the proportion of heterozygous loci in an individual organism, which is what we have been discussing, and the proportion of loci in the population at which more than one allele exists. The latter is called the **proportion of polymorphic loci** and of course is the larger of the two proportions. Usually loci with frequencies of less than 0.01 are excluded from the calculation of polymorphism, because in a large population virtually every locus has more than one allele present. But then the arbitrarily chosen cutoff point determines the proportion of polymorphic loci. The average heterozygosity does not have this troublesome arbitrariness; so I prefer to use it as a measure of variability. Another reason for our interest in heterozygosity is that, as we shall see in Chapter 4, the rate of change caused by selection is proportional to the amount of heterozygosity. If a drosophila is heterozygous at 12

TABLE 1-9

Mean heterozygosity for electrophoretically detected proteins in natural populations.

ORGANISMS	NUMBER OF SPECIES	AVERAGE HETEROZYGOSITY	STANDARD DEVIATION
Drosophila	34	0.123	0.053
Other insects	122	0.089	0.060
Crustaceans	122	0.082	0.082
Mollusks	46	0.148	0.170
Fish	183	0.051	0.035
Amphibians	61	0.067	0.058
Reptiles	75	0.055	0.047
Birds	46	0.051	0.029
Mammals	184	0.041	0.035
Dicot plants	40	0.052	0.049

Data from E. Nevo et al. 1984. *Lecture Notes in Biomathematics.* 53:13.

percent of its loci and it has 5000 loci, the average fly is heterozygous for 600 loci. It can produce 2^{600} kinds of gametes, an astronomical number.

Vertebrates seem to be less heterozygous than invertebrates, although we do not know why. Perhaps the reason is that invertebrates have larger population numbers and, as we shall discuss later, heterozygosity is less in small populations. The sample of loci tested thus far mainly concerns protein-producing regions. Biologists are just beginning to investigate systematically the amount of variability in DNA as a whole.

The most efficient way of measuring the amount of variability in all of the DNA, not just in the regions coding for proteins, is by using restriction endonucleases that recognize specific short DNA sequences and identify RFLPs. Soon data will be available for many representative organisms. We already know that the amount of heterozygosity per nucleotide is greater for the DNA as a whole than for protein-coding regions. Intervening sequences and pseudogenes have the largest amount of heterozygosity. The general picture, with a few conspicuous exceptions, is that the less significant a nucleotide difference is, the more likely it is to be heterozygous. The possible reason for this tendency will be discussed in Chapter 7, when we consider molecular evolution.

The amount of genetic diversity is very important to population biologists. How does the amount of diversity within a local population compare with that of a whole species? How does the variability within a race compare with that between races? By measuring the theoretical heterozygosity as an indication of diversity, we can compare genetic variability in populations where the random-mating assumption makes no sense — such as the *whole* human population. I shall defer discussion of this topic until after we have considered the effects of inbreeding and random gene frequency drift in Chapter 2.

A Technical Note. When comparing the amounts of genetic variability in different species, we usually prefer to use the expected heterozygosity rather than the actual amount. This is especially true for a wide-ranging species, which may be mating at random locally but not over its entire range. We therefore use the allele frequencies and compute the heterozygosity as if random mating occurred throughout the whole species; in other words, we use Equation 1-3. To emphasize that we are measuring allelic heterogeneity rather than heterozygosity per se, we use the term **allelic diversity.** The larger the number of alleles and the closer they are to equal frequency, the greater the diversity.

1-3 X-LINKED LOCI

In the female, which has two X chromosomes, the genotypic proportions of X-linked alleles in a randomly mating population are the same as those for autosomes. In the male, which has a single X chromosome, the genotypic frequencies of X-linked alleles are simply the allele frequencies. Of course in an organism where the XY sex is female, these statements are reversed.

For example, the X-linked blood group antigen, Xg^a, is found in 0.646 of human males. Since either homozygous or heterozygous females express the antigen, we would expect a frequency of $0.646^2 + 2(0.646)(0.354)$, or 0.875, in females. The observed proportion is 0.893, which is in good agreement.

The most important consequence of this difference in the two sexes occurs with rare recessive alleles. If the frequency of affected males is q, the allele frequency, then the proportion of homozygous females is q^2. Rare recessive X-linked traits are much more common in males: the rarer the trait, the greater the relative excess of males affected.

A familiar example of a human X-linked trait is color blindness. Table 1-10 gives the data from a classic study of more than 18,000 school children in Oslo, Norway. As expected, the number of affected females is much less than that of the males, but the quantitative agreement is not good. Based on the male frequency, the expected number of females is (0.0064)(9072), or 58.1, whereas the observed number is only 40.

Is the difference statistically significant? The allele frequency, as estimated from males, is simply $q = 0.0801$; from females q is the square root of 0.0044, or 0.0663. How do we average these numbers to get a best estimate for a chi-square test? If we use the maximum likelihood method (see Appendix 4), the best estimate of q is 0.0772. (This figure is closer to the male estimate, as expected, because more males are

TABLE 1-10

Frequencies of color blindness in Norwegian schoolchildren.

	MALES		FEMALES		
	NUMBER	PROPORTION	NUMBER	PROPORTION	EXPECTED
Color Blind	725	$0.0801 = q$	40	0.0044	$q^2 = 0.0064$
Normal	8324	0.9199	9032	0.9956	
Total	9049		9072		

Data from G. Waaler. 1927. *Zeit. ind. Abst. Vererb.* 45:279.

affected and also because the male ratio is a direct measure of the allele frequency whereas the female ratio gives only an indirect estimate.) Using the maximum likelihood estimate of q to get expected numbers in both sexes leads to a chi-square value of 4.77 for one degree of freedom, giving a probability of 0.03. The results differ significantly from the expectations.

We can determine the cause of the discrepancy by examining the phenotypes more carefully and distinguishing between those with a green deficiency and those with a red deficiency. If we designate red deficient by r, green deficient by g, and normal by $+$, the three kinds of X chromosomes are $r+$, $+g$, and $++$. Assume that the two loci are complementary; that is, a female of genotype $+g/r+$ has normal vision. Then, as shown in Table 1-11, the data are internally consistent.

Note that the expected proportion of $+g/r+$ females is $2p_{+g}p_{r+} = 0.0023$, accounting for the discrepancy in the expected number of affected females between the simpler and the more exact hypotheses. I have gone through this rather elaborate example to illustrate that allele frequency analysis is often quite powerful. Despite there being no pedigree information — the data came solely from the measurement of different kinds of color blindness in schoolchildren — the analysis has suggested the

TABLE 1-11

A more detailed breakdown of the data in Table 1-9.

	MALES		FEMALES		
	NUMBER	PROPORTION	NUMBER	PROPORTION	EXPECTED
Red deficient	174	$0.0192 = p_{r+}$	3	0.0003	$p_{r+}^2 = 0.0003$
Green deficient	551	$0.0609 = p_{+g}$	37	0.0041	$p_{+g}^2 = 0.0037$
Normal	8324	$0.9199 = p_{++}$	9032	0.9956	
Total	9049		9072		

two-locus hypothesis and has provided evidence for its validity. Gene frequency analysis is one of the most powerful techniques for studying the genetics of humans or other nonlaboratory species where experimental matings cannot be made. This survey was first published in 1927; it still stands as one of the best.

Direct supporting evidence for the two-locus hypothesis has come from family studies. The most interesting prediction of the hypothesis is that some normal women ($+g/r+$) should produce both kinds of color-blind sons; this situation has been found. Our analysis says nothing about how close the two loci are to each other, for the analysis deals with X chromosomes as units. Crossover sons from doubly heterozygous women have been reported. As expected, some males are *rg* and hence deficient in both red and green vision. They are very rare and none seem to have been present in the Norwegian school sample.

1-4 THE ISOLATE EFFECT

In human and other populations, rare alleles often occur at high frequencies in local populations. An island, an isolated area, or a small sect may have a high incidence of some genetic trait that is rare in the general population; or the trait may be completely absent. The high frequency of some alleles and the absence of others implies that the local population traces back to a small number of ancestors.

In most of the world such human **isolates** are being broken up by migration. This trend is especially true in the United States, which is an amalgamation of previously isolated populations from throughout the world and has a great deal of internal migration. All this mixing of previously isolated groups has an important genetic consequence: a decreased frequency of recessive traits.

Suppose that in one local population the frequency of the recessive allele for albinism is q_1, whereas in another population of the same size the frequency is q_2. Now assume that, because of improved transportation, the two communities become one randomly mating unit. What will this change do to the incidence of albinism?

Before the amalgamation the average incidence of albinism was $(q_1^2 + q_2^2)/2$. Afterward the average allele frequency, designated by $\bar{q}$, is $(q_1 + q_2)/2$, and the incidence is the square of this amount. We can easily show that the average incidence of the trait is greater before fusion than after. Let the allele frequencies in the two populations be $\bar{q} + x$ and $\bar{q} - x$. Then the average incidence before fusion was $[(\bar{q} + x)^2 + (\bar{q} - x)^2]/2 = \bar{q}^2 + x^2$. Since x^2 cannot be negative, the proportion of albinos is always less after fusion than before (unless the two populations had the same allele frequency).

Although I shall not prove it here, this conclusion is true regardless of whether or not the two populations are of equal size. To whatever extent the isolates are broken up by migration, the average proportion of homozygotes is reduced. For a more general treatment of this principle, see Crow and Kimura, hereinafter referred to as C&K, (1970), pages 54–55.

Although I know of no accurate data extending over a long enough time period to demonstrate such a trend, the frequency of persons now affected by rare recessive diseases must surely be less than it was a few generations ago. Thus one sanguine effect of increased population mobility has been a reduction of recessive disease. Also, to the extent that recessive alleles are responsible for general weakness and poor health, these conditions have been ameliorated by isolate breaking. I do not intend, however, to imply that isolate breaking accounts for more than a small fraction of the improved health and life expectancy that we have enjoyed during the past century. Most of these benefits are undoubtedly due to an improved environment and control of infectious diseases.

We just saw that if a population comprises several geographical groups, each mating at random, the whole population will not exhibit Hardy-Weinberg ratios: the proportion of homozygotes will be somewhat greater than the square of the average allele frequency, and heterozygotes will be correspondingly less frequent. The same situation occurs in a population where the allele frequencies vary among different age groups, as might happen if migrants tend to be of a particular age. Age structure can have the same effect as geographical structure in causing departures from Hardy-Weinberg ratios.

When we consider the complicated structure of real populations of most species, including our own, it is surprising to find that such simple assumptions as those leading to the Hardy-Weinberg ratios are ever realistic. Yet time and again we discover that populations conform quite closely to these ratios. When they don't, finding out why not is a challenge. Absence of Hardy-Weinberg ratios is often the first clue to an interesting finding, such as those we saw in Tables 1-6 and 1-10.

1-5 RANDOM MATING WITH TWO LOCI

When we considered a single locus (or treated a larger group, such as a linked group or a chromosome, as a single unit), we found two important random-mating principles:

1. **The probability of a zygote is the product of the probabilities of the component gametes, multiplied by 2 if the combination can arise in two ways, as with heterozygotes.**
2. **The Hardy-Weinberg equilibrium is reached in a single generation after random mating starts.**

When we look at two or more loci within a gamete, we find that the first principle remains true, but not the second.

I shall use a capital letter for the gamete frequency and a lowercase letter for the frequency of alleles within the gamete. At equilibrium the frequency of the *AB* gamete, P_{AB}, is the product of the frequencies of the two component alleles, $p_A p_B$. When this relationship is true for all alleles at the two loci, we say that the population is in **gametic equilibrium,** also called **linkage equilibrium.** The latter term is a misnomer because the equilibrium applies to independent as well as linked genes; unlinked loci can be in linkage disequilibrium. Despite these semantic reservations, I shall yield to convention and use such misnomers as linkage equilibrium and linkage disequilibrium for unlinked genes.

As a specific example, assume that two alleles are at each of the two loci and designate the frequencies of the four gametes, *AB*, *Ab*, *aB*, and *ab*, as P_{AB}, P_{Ab}, P_{aB}, and P_{ab}. With random mating the 10 zygote types are in the proportions given in Table 1-12. Since the last two genotypes are usually indistinguishable, the total proportion of double heterozygotes, *Aa Bb*, is $2P_{AB}P_{ab} + 2P_{Ab}P_{aB}$.

We should expect that the population will eventually come to equilibrium, with the frequency of each chromosome equal to the product of the frequencies of its constituent genes; for example, $P_{AB} = p_A p_B$. If the population starts at some other value, recombination will bring it toward the equilibrium. However, this process may be slow for two reasons: (1) the genes may be closely linked, so that recombinants are rare; and (2) recombination has a genetic consequence only in double heterozygotes; for all other genotypes the output gametes are the same as the input gametes regardless of

TABLE 1-12

Frequencies of the 10 zygotic genotypes under random mating. P_{AB}, P_{Ab}, P_{aB}, and P_{ab} are the frequencies of the four gametes *AB*, *Ab*, *aB*, and *ab*. Homozygotes are in the first column, single heterozygotes in the second, and double heterozygotes in the third. Ordinarily the two genotypes in the third column are phenotypically indistinguishable.

HOMOZYGOTES		SINGLE HETEROZYGOTES		DOUBLE HETEROZYGOTES	
$\frac{A\ B}{A\ B}$	P^2_{AB}	$\frac{A\ B}{A\ b}$	$2P_{AB}P_{Ab}$	$\frac{A\ B}{a\ b}$	$2P_{AB}P_{ab}$
$\frac{A\ b}{A\ b}$	P^2_{Ab}	$\frac{A\ B}{a\ B}$	$2P_{AB}P_{aB}$	$\frac{A\ b}{a\ B}$	$2P_{Ab}P_{aB}$
$\frac{a\ B}{a\ B}$	P^2_{aB}	$\frac{A\ b}{a\ b}$	$2P_{Ab}P_{ab}$		
$\frac{a\ b}{a\ b}$	P^2_{ab}	$\frac{a\ B}{a\ b}$	$2P_{aB}P_{ab}$		

whether or not recombination occurs. Thus even if the loci are far apart on the chromosome or on independent chromosomes, the approach to gametic equilibrium is gradual because it is limited by the number of double heterozygotes, which with random mating can never exceed 1/2 and is usually less.

Now that we have a qualitative picture of the approach to equilibrium let us examine it quantitatively. The population starts with specified chromosome frequencies at generation (time) zero ($t = 0$). We look at the population at time t and ask what the proportions of different chromosomes are. We designate the frequency of the AB gamete at generation t (that is, in the gametes produced by generation t) as $P_{AB,t}$; the other gametes are represented in the same manner. Let r be the probability that a gamete is recombinant. Unlinked genes will be treated as a special case when $r = 1/2$.

An AB gamete can arise in two ways: (1) It can be nonrecombinant, with probability $1 - r$, in which case it has the same probability of being AB as do the gametes of the previous generation, which is $P_{AB,t-1}$. (2) It can be recombinant, with probability r, in which case the A and B alleles came from different parents. These two ways of gametic formation are illustrated in Figure 1-2. Since mating is at random, the probability that the allele at the first locus is A and the allele at the second locus is B is simply the product of the two allele frequencies, $p_A p_B$. We do not need to put a time subscript on

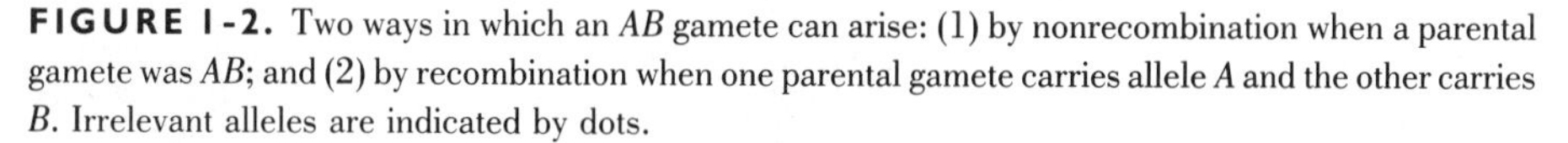

FIGURE 1-2. Two ways in which an AB gamete can arise: (1) by nonrecombination when a parental gamete was AB; and (2) by recombination when one parental gamete carries allele A and the other carries B. Irrelevant alleles are indicated by dots.

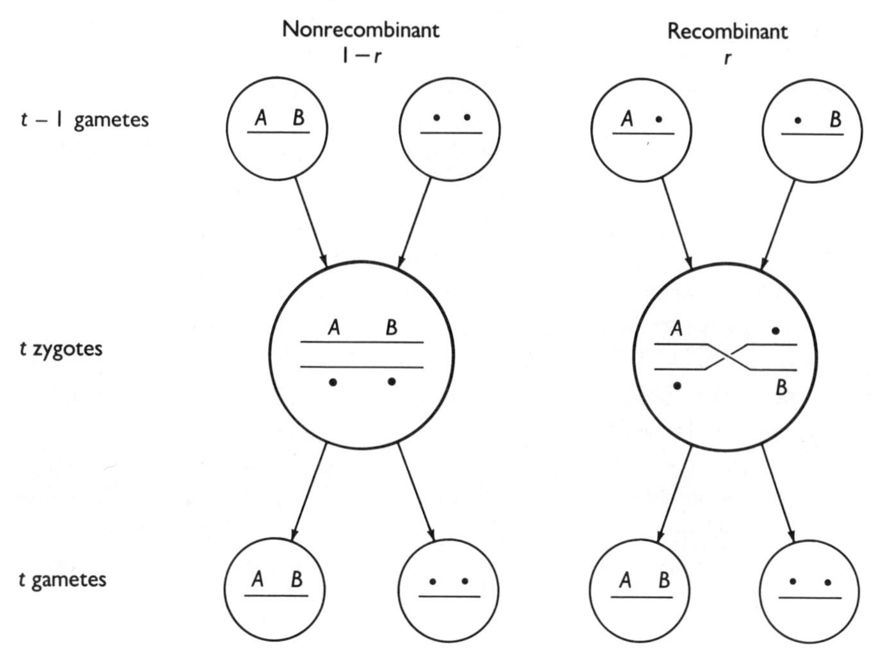

the allele frequencies, since we are assuming no selection and are ignoring random fluctuations. Thus the allele frequencies remain constant.

Putting these two probabilities together, we can write the frequency of AB gametes at time t in terms of the frequency in the previous generation, $t - 1$.

$$P_{AB,t} = (1 - r)P_{AB,t-1} + rp_Ap_B. \tag{1-4}$$

The interpretation of the formula becomes more transparent if we subtract p_Ap_B from each side and do a bit of algebraic rearranging, leading to

$$P_{AB,t} - p_Ap_B = (1 - r)(P_{AB,t-1} - p_Ap_B). \tag{1-5}$$

Note that we can go back one generation by replacing t by $t - 1$ and $t - 1$ by $t - 2$ in Equation 1-5. Making this replacement and substituting it into the rightmost expression in Equation 1-5 gives

$$P_{AB,t} - p_Ap_B = (1 - r)^2(P_{AB,t-2} - p_Ap_B).$$

If we continue this process t times, we arrive at

$$P_{AB,t} - p_Ap_B = (1 - \mathrm{r})^t(P_{AB,0} - p_Ap_B). \tag{1-6}$$

Equations 1-5 and 1-6 tell us what we wanted to know. In each generation the amount of gametic disequilibrium, measured by $P_{AB} - p_Ap_B$, is reduced by a factor equal to the amount of recombination, r, between the two loci. As t gets larger, the frequency of a gamete approaches the product of its component alleles. Although we considered only one allele at each locus, the same equation applies to all other alleles. Therefore each pair of alleles at the two loci approaches equilibrium at the same rate.

Genes that are far apart on the chromosome come to approximate linkage equilibrium rapidly, as do genes on nonhomologous chromosomes. In the latter case the rate of approach is 50 percent each generation, so that the amount of disequilibrium after five generations is only 1/32 of its original amount. On the other hand, genes less than a centimorgan apart may remain far from equilibrium for hundreds of generations.

When the population has reached equilibrium under random mating, the frequency of any genotype is the product of the frequencies at each locus. For example, the genotype *Aa bb* arises from the combination of an *Ab* and an *ab* gamete. Therefore the frequency of this genotype is $2P_{Ab}P_{ab}$, or $2(p_Ap_b)(p_ap_b)$. But this frequency is the same as $(2p_Ap_a)(p_b^2)$, which is the product of the Hardy-Weinberg frequencies at each locus. As another example the frequency of *Aa Bb* at equilibrium is $4p_Ap_ap_Bp_b$.

This principle can be extended to any number of loci. At equilibrium the frequency of any gametic type is the product of the component allele frequencies, and the frequency of any zygotic type is the product of the zygote frequencies at each locus. The rate of approach to equilibrium for three or more loci is also known, but not covered in this book. For a derivation of the three-locus case, see C&K, pages 50–52.

What is the practical value of this theory? For one thing it can tell us something about the history of a population. Since zygotic equilibrium is attained in a single generation of random mating, zygotic frequencies at individual loci can tell us nothing about the population history further than one generation back. But gametic frequencies can. For example, a survey of blood group and serum protein frequencies in Tecumseh, Michigan, showed Hardy-Weinberg ratios for each of the loci. Yet the gametic frequencies were often in disequilibrium, even for loci on independent chromosomes. Clearly the population has not been mating at random for very long, although it has for at least one generation. The explanation is that this city was populated by migrants from several countries with different allele frequencies, and the gametic types have not yet had time to come to gametic equilibrium.

The best-studied example of linkage disequilibrium is the human HLA chromosome region, which determines histocompatibility reactions. For example, the frequencies of alleles A_1 and A_2 are 0.142 and 0.304, and the frequencies of alleles B_5 and B_7 are 0.052 and 0.125. The chromosome A_1B_5 has a frequency of 0.003, whereas the product of the two alleles is $(0.142)(0.052) = 0.007$.

The recombination between these two HLA loci is just about 0.01; so about 69 generations would be required to go halfway to equilibrium, which is some 2000 years. Considering the amount of migration over the world during that time and the differential proliferation of various groups, it is no surprise that the populations are far from equilibrium for many gametic types.

Also, natural selection may favor certain chromosome combinations. However, determination of whether the observed patterns are the result of natural selection for specific linked combinations or are simply relics of an earlier time when the population was subdivided is difficult. The evidence for natural selection is good: certain HLA alleles are strongly associated with specific diseases. But whether this particular allele or an allele closely linked to it is selected is usually impossible to determine. Something in the HLA region influences the incidence of various diseases, but exactly what is not yet known. Population analysis can lead to the discovery of such associations, but molecular dissection of the chromosome region is more likely to provide a causal understanding.

Close linkage and linkage disequilibrium can be important in genetic counseling. The discovery of RFLPs very close to an allele causing a serious disease has been particularly valuable. With this knowledge, a combination of family data with information on the frequencies of different linkage associations in the population often permits more precise genetic predictions. In some cases the RFLP can be detected in early embryos, when the disease-causing gene cannot be.

1-6
USE OF ALLELE FREQUENCY ANALYSIS TO TEST HYPOTHESES

If common polymorphic traits are at all complex, pedigree analysis is usually not very revealing. Often a population analysis can provide better evidence as to the mode of inheritance. The inheritance of the major blood groups, such as the Rh system, was worked out by allele frequency analysis. More complicated systems, such as HLA, were done the same way. Maximum likelihood estimates of the allele and haplotype frequencies are now widely used in paternity testing and other medico-legal situations. The details are usually difficult and require a great deal of statistical knowledge. Therefore I shall illustrate the general principle with an already familiar and very old example — the ABO blood groups.

The ABO blood groups were discovered around 1900. It was quickly realized that they were inherited, but the exact mode of inheritance was not proven until 1924. The two competing hypotheses were: (1) two independent loci, with a dominant allele at one locus determining the A antigen and a dominant allele at the other locus determining the B antigen; and (2) three alleles at a single locus or, possibly, a closely linked cluster of alleles. Table 1-13 gives the expected proportions of the different phenotypes according to the two hypotheses.

Consider hypothesis 1 first. With random mating and gametic equilibrium, the expected proportions of zygotic types are the products of the genotype frequencies at the locus. Thus the expected proportion of group O, genotype *aa bb* on this hypothesis, is $p_a^2 p_b^2$. The expected proportion of group A, which can be either *AA bb* or *Aa bb*, is $p_A^2 p_b^2 + 2p_A p_a p_b^2$. Since $p_A^2 + 2p_A p_a = 1 - p_a^2$, the expected proportion of group A is $(1 - p_a^2) p_b^2$. These values and those of the other blood groups are given in the table.

To test whether the observed numbers are consistent with hypothesis 1, we need to know the allele frequencies. Notice that if we add the expected frequencies of groups O

TABLE 1-13

Expected proportions of ABO blood groups according to each of two hypotheses.

GROUP	GENOTYPE: HYPOTHESIS 1	GENOTYPE: HYPOTHESIS 2	EXPECTED FREQUENCY: HYPOTHESIS 1	EXPECTED FREQUENCY: HYPOTHESIS 2	OBSERVED NUMBER
O	*aa bb*	*OO*	$p_a^2 p_b^2$	p_O^2	88
A	*A– bb*	*AA, AO*	$(1 - p_a^2)p_b^2$	$p_A^2 + 2p_A p_O$	44
B	*aa B–*	*BB, BO*	$p_a^2(1 - p_b^2)$	$p_B^2 + 2p_B p_O$	27
AB	*A– B–*	*AB*	$(1 - p_a^2)(1 - p_b^2)$	$2p_A p_B$	4
Total			1	1	163

Data from Cavalli-Sforza and Bodmer. 1971.

and A, we get $p_a^2 p_b^2 + (1 - p_a^2)p_b^2 = p_b^2$. Therefore we can estimate p_b as the square root of the sum of the proportions of A and O. Thus

$$p_b^2 = \frac{88 + 44}{163} = 0.810 \qquad p_b = 0.900$$

$$p_a^2 = \frac{88 + 27}{163} = 0.706 \qquad p_a = 0.840.$$

The expected *number* of group O, for example, is $163(0.810)(0.706) = 93.21$. Table 1-14 gives this value and the other expected numbers, which are computed similarly.

The expected numbers under hypothesis 2 can be determined in a similar way. Notice that the proportion of group O plus the proportion of group A is expected to be

$$p_O^2 + p_A^2 + 2p_A p_O = (p_O + p_A)^2 = (1 - p_B)^2$$

since $p_O + p_A + p_B = 1$. Thus

$$(1 - p_B)^2 = \frac{88 + 44}{163} = 0.810 \qquad 1 - p_B = 0.900 \qquad p_B = 0.100$$

$$(1 - p_A)^2 = \frac{88 + 27}{163} = 0.706 \qquad 1 - p_A = 0.840 \qquad p_A = 0.160.$$

TABLE 1-14

The observed frequencies of ABO blood groups in African pygmies and their expected numbers according to two hypotheses.

GROUP	OBSERVED NUMBER	EXPECTED NUMBER HYPOTHESIS 1	EXPECTED NUMBER HYPOTHESIS 2	
O	88	93.21	89.26	For hypothesis 1: $\chi_1^2 = 5.06$
A	44	38.82	42.77	$P = 0.025$
B	27	21.86	25.75	For hypothesis 2: $\chi_1^2 = 0.40$
AB	4	9.11	5.22	$P = 0.53$
Total	163	163.00	153.00	

Note that the number of degrees of freedom in each case is 1. There are four classes, therefore three degrees of freedom initially. But in each case two independent quantities were estimated from the data, p_a and p_b for hypothesis 1 and p_A and p_B for hypothesis 2. Hence the number of degrees of freedom is $3 - 2 = 1$.

Since the three allele frequencies must add up to 1, $p_O = 1 - 0.100 - 0.160 = 0.740$. Table 1-14 gives the expected numbers for hypothesis 2.

The observed and expected numbers agree satisfactorily with hypothesis 2 but depart significantly from the expectations based on hypothesis 1. These data by themselves are hardly sufficient to prove one hypothesis or the other, but clearly the one-locus (or tightly-linked pair) hypothesis is favored. That a sample of 163 persons without pedigree information could provide such strong evidence shows the power of the allele frequency method in discriminating among hypotheses. Abundant evidence from all parts of the world, where allele frequencies differ greatly, now suggests that the one-locus hypothesis is correct. Pedigree information also supports this hypothesis. A few cases of *A* and *B* alleles in cis phase have been reported, which argues that there are two adjacent loci. If there really are two loci, the linkage must be very tight; the *AB* chromosome is so rare as to be unimportant statistically.

A Technical Note. For the multiple-allele hypothesis, we estimated the *A* and *B* allele frequencies from the data and computed the frequency of *O* by subtraction. We could have as easily estimated *A* and *O* and determined *B* by subtraction. Actually for these data which method we use makes little difference. My choice was based on the rarity of the AB class. Some methods of decision are less arbitrary, however. The most widely used method is the method of maximum likelihood (see Appendix 4). The expected numbers by this method are 89.1, 42.9, 25.8, and 5.2, essentially the same numbers we obtained by the crude procedure. In this case the much more difficult maximum likelihood calculations are hardly worth the effort, but in critical cases the best methods should be used. Computer routines are available for solutions to maximum likelihood equations involving simultaneous estimation of several allele frequencies.

QUESTIONS AND PROBLEMS

Unless specific information is given to the contrary, assume that the population is in Hardy-Weinberg proportions.

1. In a population there are 10 times as many *MN* as *NN* genotypes. What is the frequency of the *N* allele?
2. In a population 42 percent of the individuals are *MN*. (a) What is the frequency of the *M* allele? (b) If this question cannot be answered, what is the frequency if you are told that *M* is more common than *N*?
3. From population data we could have estimated p_M by taking the square root of the frequency of *MM* genotypes instead of by counting genes. Why is the latter method preferred?
4. Under what circumstances are Hardy-Weinberg ratios not attained in a single generation of random mating?

5. What would you infer about the history of a population if it is in Hardy-Weinberg ratios but not in gametic equilibrium for unlinked loci?

6. What proportion of *MN* children have *MN* mothers?

7. What proportion of *MM* children have two *MM* parents?

8. If x is the proportion of *aa* children in many sibships in which one parent is *aa* and the other $A-$, what is the proportion when both parents are $A-$?

9. In some varieties of sheep the presence of horns is determined by an allele that is dominant in males but recessive in females. If 96 percent of males are horned, what proportion of females have horns?

10. Show that when two loci have complete dominance and are at linkage equilibrium, the product of the double-dominant and double-recessive phenotypes equals the product of the two single dominants. You might try this calculation on the ABO blood group data in Table 1-13 as a test of the two-locus hypothesis.

11. The frequencies of chromosomes *AB* and *ab* are both 0.35, whereas those of *Ab* and *aB* are both 0.15. The recombination rate, r, between these loci is 0.2. (a) What will the frequency of *AB* be in the next generation? (b) What will the frequency of genotype *AB*/*Ab* be in the next generation? (c) What will the frequencies of *AB*/*ab* be at equilibrium? (d) What will the frequency of double heterozygotes *Aa Bb* be at equilibrium?

12. When there are only two alleles at each of two loci, linkage disequilibrium is often measured by $D = P_{AB}P_{ab} - P_{Ab}P_{aB}$. Show that D decreases by a fraction r, the recombination rate, each generation.

13. In an experiment 1158 *Drosophila* chromosomes were classified for alpha-glycerol-3-phosphate dehydrogenase-1 (Gpdh), alpha amylase (Amy), and for the Nova Scotia inversion. The following table shows the results, where F and S stand for fast and slow migration in an electrophoretic field and + and − indicate presence and absence of the inversion.

Gpdh	Amy	INVERSION	NUMBER
F	F	−	726
F	F	+	90
F	S	−	111
F	S	+	1
S	F	−	172
S	F	+	32
S	S	−	26
S	S	+	0

(a) What is the heterozygosity for Gpdh? (b) Calculate D for each of the three pairs.

14. In England one study showed that 0.133 men were bald. Among bald men, 0.56 had bald fathers. Are these data more consistent with baldness being X linked, Y linked, autosomal dominant, or autosomal recessive (the latter two being expressed only in males)?

15. If the frequencies of red- and green-deficient males are 0.02 and 0.06, what proportion of males would be totally color blind at linkage equilibrium?

16. Assume that the amount of recombination between the two loci in Problem 15 is 0.05. At what rate is linkage equilibrium approached?

17. Two equal-sized populations, 1 and 2, have frequencies q_1 and q_2 of the recessive allele a. The populations are fused into a single, randomly mating unit. (a) What is the proportion of aa homozygotes in the mixed population? (b) What is the answer to (a) if population 1 is four times as large as population 2?

18. Two loci are each segregating for two alleles: A, a and B, b. $P(AB) = 0.35$, $p(A) = 0.7$, and $p(B) = 0.6$. (a) What are the frequencies of the nine genotypes if the previous generation mated at random? (b) Is this population in linkage equilibrium? (c) If not, compute D (see Problem 12).

19. When two inbred lines, $aa\ bb$ and $AA\ BB$, where the two loci are on nonhomologous chromosomes, are crossed and the F_1 are self-fertilized, gametic equilibrium is attained in the next generation. Why doesn't Equation 1-6 give this result?

20. (a) Show that $(x^2 + y^2)/2 > [(x + y)/2]^2$. (b) How is this related to isolate breaking?

21. Use the MN data in Table 1-2 to verify the isolate principle. (Compute the average heterozygosity in the existing populations, assuming all populations are of equal size. Then compute the value if these equally sized populations were mixed.)

22. With n alleles at a locus, how many genotypes are there?

23. Is the rate of approach to linkage equilibrium the same for two loci at opposite ends of a long chromosome and two unlinked loci?

24. If P, Q, and R stand for the proportions of AA, Aa, and aa genotypes, respectively, the quantity PR/Q^2 is often used to test Hardy-Weinberg proportions. (a) What is the value of this quantity in a Hardy-Weinberg population? (b) Will the value increase or decrease when inbreeding occurs?

25. Among human X chromosomes the following proportions represent four conditions: A, normal; B, recessive allele for red deficiency; C, recessive allele for green deficiency; and D, both recessive alleles. (a) What fraction of women would have normal vision but carry both color-deficiency alleles? (b) Assuming no recombination, what fraction of women would be normal but have sons who are all color blind?

CHAPTER 2

INBREEDING, RANDOM DRIFT, AND ASSORTATIVE MATING

Departures from random mating usually occur because of inbreeding or assortative mating, or both. **Inbreeding** occurs when mates are more closely related than if they were chosen at random. **Assortative mating** occurs when mates resemble each other phenotypically. Among humans, deaf people tend to marry each other, as do those with tuberculosis. Husbands and wives are correlated for skin color, height, and IQ. As you know, related individuals have similar genotypes. Since individuals with similar phenotypes tend to have similar genotypes also, we expect that inbreeding and assortative mating will have similar consequences, which they do.

In natural populations we can often find small amounts of inbreeding and assortative mating. It is not always clear which is cause and which effect, or whether there is a third cause of both. For example, if animals from the same locality mate simply because of proximity, they are likely to resemble each other genetically as well as phenotypically.

Finally, random changes in allele frequencies occur from generation to generation in all finite populations, especially in small populations, and produce consequences rather similar to those of inbreeding. The result of these changes is called **random genetic drift,** or the "Sewall Wright effect," after the man who emphasized the possible evolutionary consequences of such changes.

2-1 INBREEDING

The phenotypic consequences of inbreeding are well known and were observed long before Mendelian inheritance was understood. Charles Darwin wrote a long book, *The Effects of Cross- and Self-Fertilization in the Vegetable Kingdom,* which provides a full and accurate description of inbreeding effects. Except for his lack of information on the causes of these effects, the book reads well even today. The consequences of inbreeding are (1) reduced vigor and fertility and (2) uniformity and "true breeding" within inbred lines.

Repeated self-fertilization or sib mating produces strains that are almost completely homozygous. Such strains have been invaluable for genetic and medical research. For example, since all mice within an inbred strain are essentially identical, they tolerate skin and organ transplants from one another. The homozygosity has a price, however. Highly homozygous strains in almost every outbreeding species are characterized by lowered vigor, poor survival, and reduced fertility. An experimenter who wants a strain that is uniform but still high in vigor can cross two inbred lines. The progeny are genetically identical but heterozygous for all the alleles that differ in the two strains.

The main reason why homozygosity reduces vigor is that most recessive alleles in the population are harmful. A heterozygous locus becomes homozygous in one of two ways: (1) homozygous for the dominant allele, which changes nothing phenotypically, or (2) homozygous for the recessive allele, which expresses the harmful effects of the allele. Thus inbreeding has a deleterious effect by uncovering harmful recessive alleles that were previously concealed by heterozygosity with dominant alleles.

Why are recessive alleles harmful? The underlying reason is that most mutant alleles are harmful. Dominant mutants are quickly eliminated by natural selection, but recessive mutants are protected from natural selection by being hidden in heterozygotes and therefore persist in the population.

A second reason why inbreeding is harmful is that at some loci the heterozygote is superior in fitness to either homozygote. By decreasing the number of heterozygotes, inbreeding decreases fitness. Very few such loci are known, and this effect is generally considered to be a relatively unimportant cause of inbreeding decline.

We can easily deduce the quantitative consequences of the most extreme form of inbreeding — self-fertilization. Suppose we start with a population in which a fraction

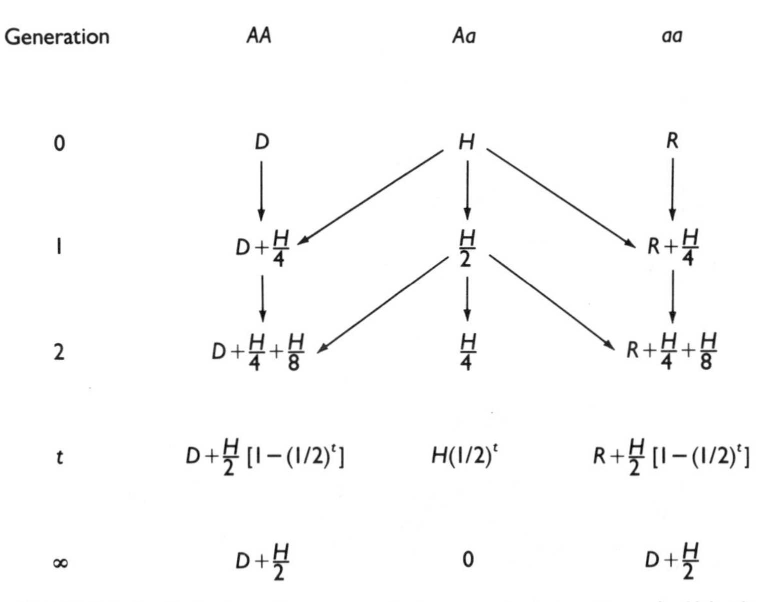

FIGURE 2-1. Reduction of heterozygosity in successive generations of self-fertilization.

D are homozygous dominant *(AA)*, *H* are heterozygous *(Aa)*, and *R* are homozygous recessive *(aa)*. Each homozygous genotype produces only its own genotype when self-fertilized. The progeny of heterozygotes are $1/4$ *AA*, $1/2$ *Aa*, and $1/4$ *aa*; so $1/4$ of the progeny of the heterozygotes is added to each of the homozygous classes. This mating system is illustrated in Figure 2-1.

The proportion of heterozygotes is reduced to half of its previous value each generation, and the remaining half is divided equally between the two homozygous classes. Note that the 50 percent reduction of heterozygotes each generation is independent of the initial proportion and of whether or not the population started in Hardy-Weinberg ratios.

We would expect that less extreme inbreeding would lead to qualitatively similar, but less extreme, results—a decrease in heterozygosity but at a slower rate. Our expectations are correct. For example, the frequencies of heterozygotes in successive generations of sib mating are in the ratios 1, $2/2$, $3/4$, $5/8$, $8/16$, $13/32$, and so on, as shown in Figure 2-2. The denominator doubles each generation, whereas the numerator is given by the Fibonacci sequence (each number is the sum of the two preceding numbers).

We can compute these fractions by laboriously keeping track of the numbers from generation to generation until we discover a general rule. Indeed, working out the consequences of different inbreeding systems was a popular game among geneticists in the 1910s. This activity stopped abruptly in 1921 when Sewall Wright, then at the United States Department of Agriculture (and at the time of this writing still active at the University of Wisconsin), discovered a simple algorithm by which the change of

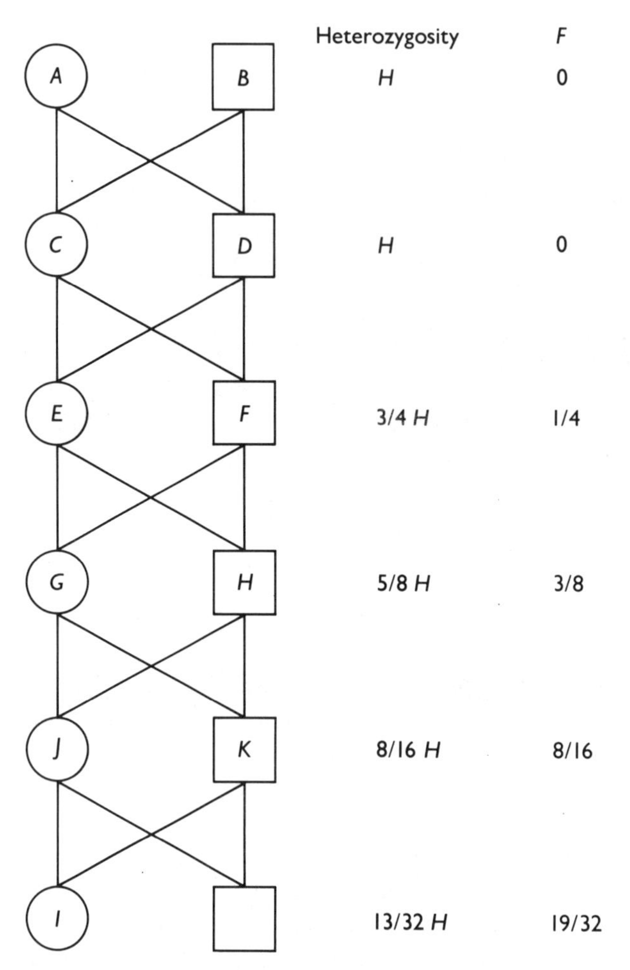

FIGURE 2-2. Reduction of heterozygosity in successive generations of sib mating. The column headed F will be explained later in this chapter.

heterozygosity can be measured for any pedigree, however complex. Wright's original derivation was based on correlations, and many find it hard to understand. In the next section I shall use a simpler, probabilistic approach.

Kinship and Inbreeding Coefficients. Since the inbreeding of an offspring is a consequence of the relatedness of its parents, we might expect to measure kinship and inbreeding similarly. If two individuals are related, they have one or more ancestors in common. The standard measure is the **coefficient of kinship,** also called the coefficient of consanguinity or coefficient of coancestry. The coefficient of relationship, a logical term for this measure, has been preempted for another quantity, which I shall discuss in Chapter 5.

The coefficient of kinship is the probability that two homologous genes, one chosen at random from each of two individuals, are "derivative", or **identical by descent.** Two genes are identical by descent if both are descended from the same gene in a common ancestor or if one gene is descended from the other. I shall use the symbol F_{JK} to represent the coefficient of kinship of two individuals, J and K. For example, the coefficient of kinship of a parent and child is 1/4: since a parent and child share one allele at an autosomal locus, the probability of drawing this allele from both parent and child is 1/4. I think you will agree that this simple definition captures the essence of a genetic relationship. In the next section I shall derive an algorithm for calculating F_{JK} when the relationship is more complicated.

We define the **coefficient of inbreeding,** F_I, of an individual as the probability that the two alleles at a locus are identical by descent. This measure is the same as the coefficient of kinship of the individual's parents, because the zygote is produced by drawing a gene at random from each parent. Hence if J and K are the parents of I, then $F_I = F_{JK}$. The use of the same letter for both the inbreeding and kinship coefficients should not cause any confusion. If there is one subscript, we are talking about gene identity within a single individual; if there are two subscripts, we are talking about two alleles in different individuals.

To see how inbreeding affects genotype frequencies, let us once again use the gene pool model. We draw one gene from the pool and then draw a second. On the second drawing, with probability F, we pick the same gene again. The probability of the first gene being allele A is p_A and of the second being identical to the first is F, so that the probability of getting two A alleles identical by descent is $p_A F$. The probability that the second gene is not identical to the first is $1 - F$, but both genes can still be A. The probability of both genes being A in this situation is p_A^2, since if the alleles are not identical they are independent. Putting these probabilities together, we have

$$\begin{aligned} P_{AA} &= p_A F + p_A^2(1 - F) \\ P_{Aa} &= 2p_A p_a(1 - F) \\ P_{aa} &= p_a F + p_a^2(1 - F). \end{aligned} \qquad \text{2-1}$$

The frequency of heterozygotes does not include a term corresponding to $p_A F$, since if the two alleles are different they cannot be identical by descent. For the present I am ignoring the possibility that one of the alleles has mutated since they descended from a common ancestral allele.

In general, regardless of the number of alleles, the probability of a homozygote is the allele frequency times F plus the square of the frequency times $1 - F$. The probability of any heterozygote is twice the product of the frequencies of the two constituent alleles times $1 - F$. If $F = 1$, the population is entirely homozygous. If $F = 0$, the population is in Hardy-Weinberg proportions. In symbols,

$$P_{ii} = p_i F + p_i^2(1 - F)$$
$$P_{ij} = p_i p_j(1 - F), \qquad 2\text{-}2$$

where P_{ij} is the frequency of the ordered genotype $A_i A_j$, or half the frequency of $A_i A_j$ heterozygotes.

Since the frequency of any heterozygous type is proportional to $1 - F$, the sum of all heterozygotes is also proportional to $1 - F$. Therefore another meaning of F, more closely related to practical use, is

***F* is the proportion by which the heterozygosity is reduced relative to the heterozygosity in a randomly mating population with the same allele frequencies.**

In symbols, the heterozygosity for a specified value of F, Het_F, is related to the heterozygosity under random mating, Het_0, by

$$Het_F = Het_0(1 - F). \qquad 2\text{-}3$$

Let us use this theory to answer a practical question. The incidence of recessive cystinuria in the children of unrelated parents is about 1/8100. What is the probability of having an affected child from an incestuous mating of father with daughter? The allele frequency, p_a, from the incidence when mating is random is the square root of the incidence, or 1/90. We noted earlier that F_{JK} for parent and child is 1/4, so this value is also the inbreeding coefficient of the child. Using Equation 2-2, we have $P_{aa} = (1/90)(1/4) + (1/90)^2(3/4) = 93/32{,}400$. The risk of an affected child is 23 times as high as if the parents were unrelated.

Notice that the rarer a recessive allele, the greater the relative increase in homozygous recessives with inbreeding. For example, if the allele frequency were half as great as in the preceding example (that is, 1/180 instead of 1/90), the risk with father-daughter inbreeding would be about 50 times that with unrelated parents. Some very rare human conditions have been found only in children of consanguineous matings. Indeed, if a mutant occurred only once, it could be homozygous only by inbreeding. On the other hand, for common recessive genes the homozygote frequency increases relatively little with inbreeding. If the gene frequency were 1/10, the increase would be about three-fold.

First-cousin matings are a much more common form of inbreeding in the human population. You will learn later in the chapter that in this case $F = 1/16$. You may wish to verify that the increase in incidence of cystinuria from first-cousin matings is about 6.6-fold.

Lest these examples give an exaggerated picture of the risks of consanguineous matings, let me note that although the relative increase is great for homozygous rare recessive alleles in consanguineous matings, such traits are so uncommon that the absolute increase is not large: a 6.6-fold increase in cystinuria changes the incidence from 1/8100 to 1/1234, which is still very small. The overall effect of a first-cousin mating is to increase the risk of early childhood death, congenital anomaly, or severe disease by about 50 percent. For example, the extensive studies in Hiroshima and Nagasaki show that congenital malformations increased by 45 percent, from 0.011 to 0.016, in first-cousin marriages. Occasional consanguineous marriages add relatively little to the incidence of disease and malformation.

Now that you know the meaning and use of inbreeding and kinship coefficients, let us see how to compute them.

Computing Inbreeding and Kinship Coefficients. Wright originally developed the inbreeding coefficient through correlation analysis. I will use a probability approach here, which most people find easier to understand. Wright's original paper is referenced at the back of the book.

Figure 2-3 shows a simple pedigree. The circles represent zygotes, the dots gametes. Recall that $F_I = F_{JK} =$ the probability that the alleles in gametes j and k are identical by descent. This relationship will be true if and only if j, d, b, a, a', c, e, and k all carry identical alleles. If $j \equiv k$ is understood to mean "j and k carry identical alleles at the locus in question," then $P(j \equiv k) = P(j \equiv d) \times P(d \equiv b) \times P(b \equiv a) \times P(a \equiv a') \times P(a' \equiv c) \times P(c \equiv e) \times P(e \equiv k)$, since each of the terms on the right-hand side of the equation is independent of the others. (The meiotic events in each individual are not influenced by those in other individuals; therefore the probabilities are independent.)

According to the rules of Mendelian inheritance, $P(j \equiv d)$ is obviously 1/2, as are all the other adjacent pairs except a and a'. To consider this pair, suppose that the two genes in zygote A (for ancestor) are labeled G and G'. Gametes a and a' can have four equally likely contents: G and G, G' and G', G and G', and G' and G. The first two cases are identical. The latter two are not, unless ancestor A is also inbred. If A is inbred, the probability of G and G' being identical is F_A, by the definition of F. Therefore $P(a \equiv a') = 1/2 + (1/2)F_A = (1/2)(1 + F_A)$. Putting these probabilities together, we get $F_I = F_{JK} = (1/2)^7(1 + F_A)$. Notice that 7 simply represents the number of individuals, J, D, B, A, C, E, and K, in the ancestral path from J to K.

If there is more than one common ancestor or more than one path through the same ancestor, we add the values for all paths. The values are added because the paths are mutually exclusive (that is, if j and k carry identical genes descended through one path, they cannot carry identical genes descended through another), and we want to know the probability that the alleles are identical regardless of which path was taken.

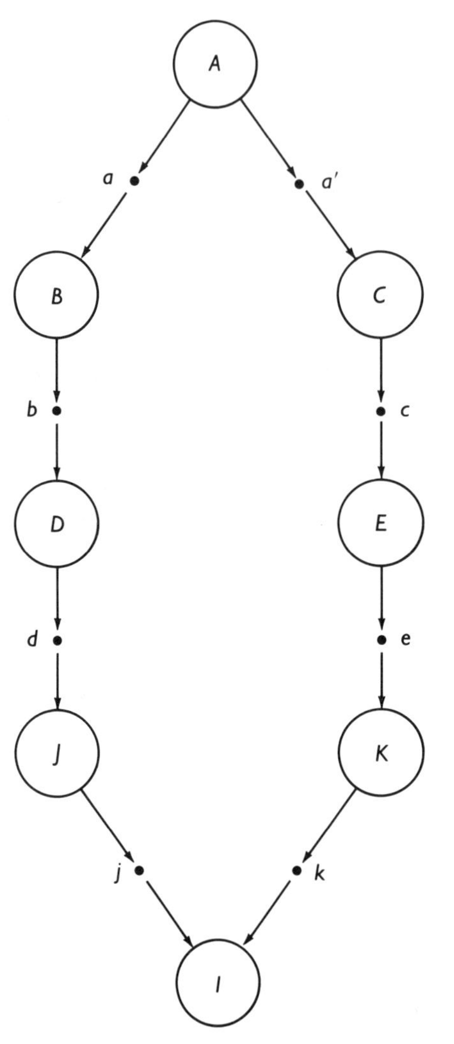

FIGURE 2-3. A simple pedigree. Circles represent zygotes; dots represent gametes. Only individuals that contribute to the inbreeding of I and the relationship of J and K are included.

Therefore the general rule that works for any pedigree, however complex, is

$$F_I = F_{JK} = \sum \left[\left(\frac{1}{2} \right)^n (1 + F_A) \right], \qquad 2\text{-}4$$

where Σ means to sum over all possible paths from J to an ancestor and back to K, n is the number of individuals in a path, and F_A is the inbreeding coefficient of the ancestor at the end of a path.

Figure 2-4 gives another example of a pedigree. Since E is inbred, we first calculate her inbreeding coefficient. The calculations are given below. The letters beneath each term indicate the relevant path, and the common ancestor in each path is underlined.

$$F_E = \underset{G\underline{M}H}{\left(\frac{1}{2}\right)^3} + \underset{G\underline{L}H}{\left(\frac{1}{2}\right)^3} = \frac{1}{4} \qquad 1 + F_E = \frac{5}{4}$$

$$F_I = \underset{J\underline{C}K}{\left(\frac{1}{2}\right)^3} + \underset{J\underline{D}K}{\left(\frac{1}{2}\right)^3} + \underset{JC\underline{E}DK}{\left(\frac{1}{2}\right)^5\left(\frac{5}{4}\right)} + \underset{JD\underline{E}CK}{\left(\frac{1}{2}\right)^5\left(\frac{5}{4}\right)} = \frac{21}{64}.$$

We can analyze any pedigree in this fashion. The only difficulty with complex pedigrees is that we must be sure that we have not overlooked any paths or counted a path twice. Animal breeders and human geneticists usually devise some systematic way of ensuring that they do not make such mistakes. Computers are better than humans at this kind of drudgery; so inbreeding coefficients for complicated pedigrees are often done by computer.

We can sometimes make complicated pedigrees more clear by indicating the direction from parent to offspring by arrows, as in Figure 2-4. The rule is to start with J (the mother, say) and go back to the ancestor against the arrows and then come back to K (the father) with the arrows, reversing the direction only once (at the ancestor) and never going through the same individual twice. Notice that although we must include the term $1 + F_A$ if the ancestor is inbred, the inbreeding of any other individual in the path is irrelevant, since the probability that an incoming and outgoing gamete from the same individual carry the same allele is $1/2$ regardless of the allele that came in through the other parent.

Figure 2-5 gives three more pedigrees, showing matings between double first cousins, uncle and niece, and half-sibs. Both the coefficient of kinship of these relatives and the coefficient of inbreeding of a child are $1/8$ in each case. Yet we have

FIGURE 2-4. A more complicated pedigree. Circles represent females; squares represent males.

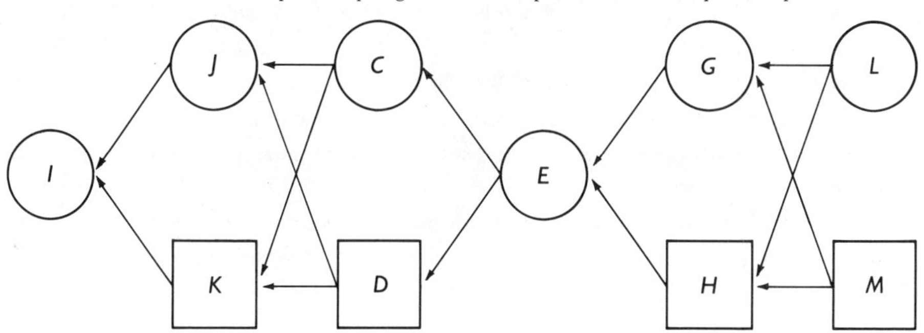

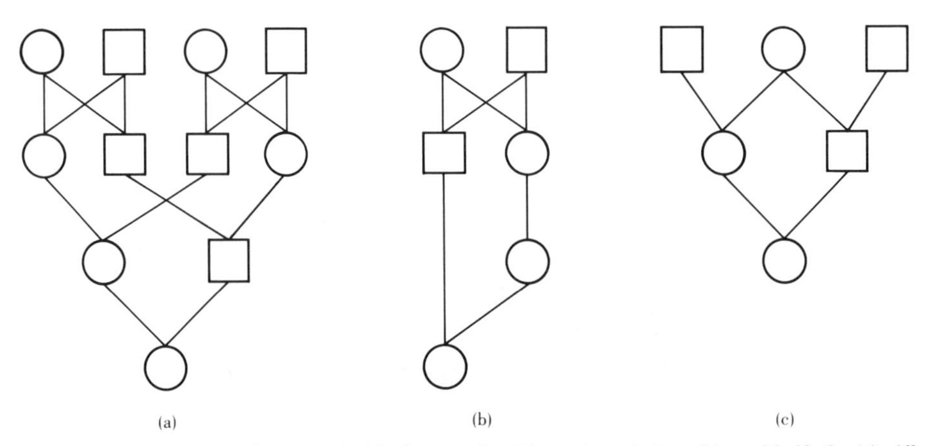

FIGURE 2-5. Matings between double first cousins (a), uncle and niece (b), and half-sibs (c). All produce an inbreeding coefficient of 1/8 in the progeny.

quite different social conventions about marriages between persons of these relationships.

You might want to test your skill in tracing paths by computing F for the first few generations of the sib-mating pedigree in Figure 2-2.

There is a conceptual difficulty about inbreeding and kinship coefficients. As we go farther back in a pedigree, we will often find more common ancestry, but each path will be longer and therefore will contribute less. Nevertheless if we could go back far enough and had complete information, the inbreeding coefficient would be very high, approaching 1. We are all related if we go back far enough, although we might have to go back to the early days of the human species. If N is the size of a randomly mating population, the average number of generations back to a common ancestral gene from which all present alleles are descended is $4N$. The derivation of this formula is beyond the scope of this book (see C&K, pages 430 ff).

There are two approaches to this difficulty. First, if we are concerned with only a few generations, a dozen or less, we will not encounter any serious difficulty. The purpose of calculating F is almost always to compare the effects of a specified degree of kinship with those of "unrelated" individuals in the same population. By unrelated we mean that the individuals have no common ancestry as far back as our pedigree knowledge extends. When we say that the effect of a cousin marriage is to reduce the heterozygosity of the child by 1/16, we mean this amount relative to persons in the same population who are unrelated as far back as the pedigrees go. We do not know the absolute heterozygosity.

The second approach, which is appropriate when we are considering long evolutionary periods, is to take mutation into account. I shall deal with this topic later in the chapter.

X-linked Loci. It makes no sense to ask about the inbreeding coefficient for the Y chromosome. Two males that are related to each other through an unbroken chain of males necessarily have the same Y chromosome; otherwise they do not. Thus the kinship coefficient is either 1 or 0. As for the X chromosome, it also makes no sense to ask about the inbreeding coefficient of a male, which has only one X. However, we may meaningfully ask for the kinship between two individuals of either sex or for the inbreeding coefficient of a female.

Two simple modifications of Wright's algorithm make it applicable to X-linked genes. First, notice that the X chromosomes in the egg that produced a male and in an X-carrying sperm from that male are necessarily identical; so the probability of this segment of the path is 1 instead of the 1/2 it would be if the locus were autosomal. Since multiplying by 1 has no effect, we simply ignore all males in the path. Likewise, there is no way for a male to transmit an X chromosome to his son; so every time two successive males occur in a path, the X chromosome chain is broken and the probability of identity through this path is 0. Hence we can state the modifications for an X-linked locus (or for any locus in haploid-diploid organisms, such as honeybees) as follows:

1. **Ignore all males in a path.**
2. **Ignore all paths with two adjacent males.**

To test your understanding of these rules, compute the inbreeding coefficient of *I* in Figure 2-4 for an X-linked locus. The answer is 13/32.

2-2 PHENOTYPIC EFFECTS OF INBREEDING

We have seen that inbreeding increases the frequency of recessive diseases and generally decreases vigor. In natural populations of *Drosophila melanogaster*, the average number of recessive lethal alleles per autosome ranges from about 0.25 to 0.40. Since there are two pairs of major autosomes and the X chromosome has very few recessive lethals, the mean number of lethals per haploid set ranges from 0.5 to 0.8. The average diploid fly thus carries 1.0 to 1.6 recessive lethal alleles. In addition, many deleterious alleles have mild homozygous effects. If we count one "lethal equivalent" as a single gene that is lethal, or two genes each with probability 1/2 of causing death, or ten genes with probability 1/10 of causing death, and so forth, we can say that the average fly also carries some 0.5 lethal equivalent from mildly deleterious mutants. Altogether, then, the average fly carries 1.5 to 2.1 lethal equivalents that would be expressed if made homozygous.

Data from laboratory animals and domestic livestock are not very helpful for such analyses, because the animals are usually quite inbred and recessive lethals have been largely eliminated. We can get some information from human consanguineous marriages, which leads to estimates similar to those for Drosophila. We humans carry an average of about two lethal equivalents per zygote, based on Japanese data in the 1950s and 1960s. This number gets smaller each generation. Studies based on consanguineous marriages in rural France in the 1930s give four to five lethal equivalents per zygote. Clearly the better environment of recent Japan has reduced the number of lethal equivalents: higher living standards, better health care, and reduction of infectious disease have permitted genotypes to survive that would have perished under the more severe environment in rural France a generation earlier. What used to be a severely deleterious recessive gene has become only mildly deleterious. No reliable recent data are available from the United States.

With such harmful effects it is not surprising that many species of animals and plants have various devices to prevent inbreeding. For one thing, having separate sexes prevents self-fertilization. In addition, species that produce both pollen and ovules on the same plant often have elaborate devices to prevent self-fertilization. In orchids, for example, the floral parts are specifically adapted to foster insects to carry pollen from one plant to another. Charles Darwin was intrigued by such devices and as early as 1862 said, "Nature abhors perpetual self-fertilization."

One of the best understood mechanisms for preventing inbreeding is found in many plant species. There are many alleles at a specific self-sterility locus, usually designated as S. Pollen carrying a specific allele will not germinate on a plant that carries this same allele. Thus, for example, a pollen carrying allele S_1 will not grow on a plant of genotype S_1S_2 but will grow on one of S_2S_3. Clearly such a system prevents self-fertilization and reduces the likelihood of fertilization of near relatives, which are likely to share alleles. A large population may have several hundred alleles at this locus. Hence unrelated plants hardly ever share the same allele. This system is very efficient for preventing self-fertilization while not reducing general fertility.

Many animals have mechanisms or behavioral patterns that prevent mating of close relatives. Almost every human society has incest taboos. One cause of these taboos may be the increased incidence of disease and deformity caused by inbreeding. Early human societies may have recognized this correlation and erected social barriers to reduce such matings. Since many primates also avoid close inbreeding, incest avoidance may have evolved by natural selection in prehuman times and later became institutionalized. Those individuals that avoided mating with close relatives had a greater chance of having viable progeny; hence genes for such behavior would have had selective value. Of course, some causes of inbreeding taboos may be unrelated to harmful biological effects, as social anthropologists and psychologists have often pointed out. Untangling social from biological causes of human inbreeding avoidance is an intriguing problem, which is not likely to be solved in the near future.

2-3 HOW CAN WE ACCOUNT FOR SELF-FERTILIZING PLANT SPECIES?

Clearly inbreeding is deleterious. I have given several examples of inbreeding-avoidance mechanisms, yet many self-fertilizing species occur in the plant kingdom. In many cases there is good evidence that these species are descended from outcrossing species. How could evolution from random mating to self-fertilization evolve in the face of the weakening effects of inbreeding? If a species has been self-fertilized for several generations, most deleterious genes would be eliminated by natural selection, provided that the species survived the period when the recessive alleles were being expressed as homozygotes. Inbred lines of mice and maize were developed in this way. However, in the process of close inbreeding many strains become extinct. Weak inbred strains can be maintained by human intervention, but how can such strains compete successfully with other, noninbreeding species?

This problem, like so many others, was first solved by the British statistician and geneticist R. A. Fisher. The following system is the easiest to visualize and the most probable. Suppose that the plant species is regularly cross-fertilized; some mechanism prevents self-fertilization. If the population size is constant, each plant, on the average, transmits one sperm and one egg. Now suppose that one plant has the capacity to fertilize its own eggs while at the same time continuing to produce pollen that is used to fertilize other plants. Since pollen is produced in great abundance, the number of progeny produced by outcrossing pollen will not be reduced appreciably. But this plant will contribute an average of three gametes to the next generation instead of two: one egg, a self-fertilizing sperm, and an outcrossing sperm. Therefore the genes causing self-fertilization should increase rapidly in the population.

On the other hand, self-fertilization leads to lethality. Therefore the genes for self-fertilization will increase only if the 50 percent excess in progeny produced is greater than the reduction in fitness of the self-fertilized progeny. Suppose that the average plant carries two recessive lethals at independent loci. If such a plant is self-fertilized, $1/16$ of the progeny will be homozygous for both lethals, $6/16$ for one or the other, and $9/16$ for neither. The number of self-fertilizing gametes contributed will be $2(9/16) = 9/8$, which is greater than 1. Hence the genes causing self-fertilization will tend to increase.

If the number of recessive alleles causing reduction of viability and fertility is small enough that self-fertilization leads to less than a 50 percent reduction in fitness, genes causing self-fertilization will tend to increase; otherwise they will not. Thus the proportion of self-fertilizers tends to increase to 100 percent or decrease to 0. In nature we would expect to find that most plant species are either predominantly selfing or predominantly outcrossing, with very few species having a mixture of both types. Natural plant species fit this expectation.

Of course, this evolutionary trend depends on the existence of heritable mechanisms for switching from outcrossing to self-fertilization. Many possible mechanisms have been described. The argument in the preceding paragraph makes one other prediction: since self-fertilization is more likely to evolve in a species with a small number of lethal and sterile equivalents, it is more probable in a species that already has some inbreeding, such as one with a very small population.

The development of self-fertilization provides an excellent illustration of the important principle that evolution does not always move in the direction of increasing fitness; that is, it is not necessarily good for the species. The evolution of self-fertilizing species occurs despite their being less fit than outcrossers. It is very unlikely that *all* the deleterious recessives are eliminated, even after several generations of self-fertilization. In addition, at some loci the heterozygote may be more fit than either homozygote. Such loci are known to exist in some species. The classic example is beta hemoglobin, in which the *AA* homozygote is susceptible to malaria and the *SS* homozygote has sickle-cell anemia, whereas the heterozygote is free of both problems. Although overdominant loci, in which the heterozygote is more fit than either homozygote, are thought to be rare, even a small minority would mean that an inbred strain could never be as fit as an outcrossing one. Thus many plant species evolving into self-fertilizers are likely to become extinct through competition with other species.

Several examples of obligatory sib-mating occur in insects. In one case the insects are parasites on other insect species. The parasite lays eggs in the host, and the progeny mature in the host and mate before they emerge. Thus all matings are between sibs. Presumably this species had a small enough number of lethal and sterile equivalents to permit evolution in this direction. Clearly this system can be advantageous to a parasitic insect that exists in small numbers and in which individuals may therefore have trouble finding mates. This benefit may be sufficient to offset the deleterious effects of inbreeding.

2-4 RANDOM GENETIC DRIFT

In a large population with random mating, two mates are very unlikely to be close relatives. On the other hand, if the population is small, and especially if it has been small for a long time, many individuals will be related. Thus random matings are likely to be consanguineous. We would expect that small populations, even if mating within the population is random, would have consequences very similar to those of inbreeding.

Rate of Decrease of Heterozygosity. To illustrate the effects of a finite population, I shall again use the gene pool device. Suppose that N diploid individuals are in the parent generation and that each of the individuals contributes equally to a large pool of gametes. To produce an offspring, we draw two genes from the pool. The probability of drawing the same parental gene twice is $1/2N$, since there were $2N$ parental genes. The probability of drawing two different genes is $1 - 1/2N$. However, the probability that the two different genes are in fact identical because of common ancestry is the inbreeding or kinship coefficient of the parent generation. Since mating is random, the inbreeding coefficient of an individual and the kinship coefficient of two different individuals are the same. (I am including the possibility of self-fertilization here, but will consider separate sexes later in the chapter.)

If we let F_t represent the inbreeding coefficient—that is, the probability of identity by descent—in the tth generation, then from the arguments in the preceding paragraph

$$F_t = \frac{1}{2N} + \left(1 - \frac{1}{2N}\right)F_{t-1}. \qquad 2\text{-}5$$

The meaning will be more transparent if we rewrite this equation as

$$1 - F_t = \left(1 - \frac{1}{2N}\right)(1 - F_{t-1}).$$

Now recall that $1 - F$ is proportional to the heterozygosity; that is, $Het_t = K(1 - F_t)$, where K is a constant. Making this substitution, and noting that the K's cancel, we have

$$Het_t = \left(1 - \frac{1}{2N_e}\right)Het_{t-1}. \qquad 2\text{-}6$$

In words, Equation 2-6 says that in a population of effective size N_e, heterozygosity is reduced each generation by a factor $1/2N_e$. This result was first obtained by Sewall Wright. Het_{t-1} is the same function of Het_{t-2} as Het_t is of Het_{t-1}. Making this substitution and continuing for t generations leads to

$$Het_t = \left(1 - \frac{1}{2N_e}\right)^t Het_0. \qquad 2\text{-}7$$

I have added the subscript e to bring out another point. The model assumes that a gene is equally likely to come from any parent. This assumption does not mean that each parent produces exactly the same number of progeny, for there will be random variability, but that each parent has the same *expected* number. In this idealized population the number of progeny per parent will have a binomial distribution. Most actual populations depart from this ideal. Therefore we define the **effective population number,** N_e, as the size of an idealized population that has the same probability of identity as the actual population being studied. Clearly the effective population number is related to the number of adults rather than that of juveniles. This distinction can be important in organisms in which the death rate is very high in early stages, such as many invertebrates. Also, the degree of fertility usually varies. Thus the effective population number is generally somewhat less than the number of adults of reproductive age.

I shall not derive the formula for heterozygosity reduction when no self-fertilization takes place. You can find the exact formula in C&K, page 102. To a very good approximation, the formula is

$$Het_t \approx \left(1 - \frac{1}{2N_e + 1}\right) Het_{t-1} \approx \left(1 - \frac{1}{2N_e + 1}\right)^t Het_0 . \qquad 2\text{-}8$$

This approximation is accurate except for the first few generations in very small populations. A severe test of the approximation is repeated sib mating, in which $N_e = 2$. From Figure 2-2 the relative heterozygosity in successive generations is 1, 3/4, 5/8, 8/16, 13/32, 21/64, 34/128, and so on. The proportion by which the heterozygosity is reduced in successive generations is 0.250, 0.167, 0.200, 0.187, 0.192, 0.190, and so forth, rapidly becoming approximately 1/5, the value given by Equation 2-8. The exact limit, for comparison, is 0.191.

Here is one more useful formula. If the numbers of the two sexes are different, then

$$\frac{1}{N_e} = \frac{1}{4}\left[\frac{1}{N_{ef}} + \frac{1}{N_{em}}\right], \qquad 2\text{-}9$$

where N_{ef} and N_{em} are the effective numbers of females and males. For example, a herd of cattle with one male and a very large number of females has an effective number of approximately 4.

Notice that in Equation 2-8 the population numbers appear in the denominators. Thus if the numbers fluctuate from time to time, the smaller numbers will be more influential. If a population goes through small bottlenecks, the mean effective population number may be much less than the simple average of the population numbers might suggest. In fact, to a good approximation the average effective population number of a population with fluctuating size is the harmonic mean of the effective numbers in successive generations.

There is one tricky concept concerning the change in heterozygosity in a finite population. Since mating is random, the genotypes remain in Hardy-Weinberg ratios. Yet the proportion of heterozygotes becomes less. How can this happen? The reason is that the allele frequencies are changing. You might think that the change is equally likely to be positive or negative, which is true. Yet, on the average, the gene frequency changes so that the quantity $2p_i p_j$ decreases each generation by $1/2N_e$, which follows from the derivation of Equation 2-6.

Perhaps I can make this seemingly paradoxical result more clear intuitively. Suppose that in a single generation the frequency of allele A changes from p to $p + x$. Then allele a changes from q to $q - x$. In the next generation the proportion of heterozygotes is

$$2(p + x)(q - x) = 2pq + 2x(q - p) - 2x^2.$$

But the average value of x is 0 if we are considering the average of several populations or the change in a single population over several generations. Thus we can ignore the middle term on the right-hand side of the equation. Since x^2 is necessarily positive, the proportion of heterozygotes regularly decreases despite the erratic behavior of the individual allele frequencies.

Two Examples. As a numerical example, suppose an island population starts with 20 shipwrecked people (in the manner of the mutineers from the *Bounty*, who, along with some Tahitians, founded the population of Pitcairn Island). To simplify matters, suppose that the proportions of men and women remain equal and that the population doubles each generation. The heterozygosity after five generations will be $(1 - 1/41)(1 - 1/81)(1 - 1/161)(1 - 1/321)(1 - 1/641) = 0.953$ times the original heterozygosity. Notice that more than half of the total decrease in heterozygosity took place in the first generation. If the population had remained at 20 instead of increasing, the heterozygosity would be $(40/41)^5$, or 0.884 of its initial value, which is about the equivalent of an uncle-niece mating. We would expect a substantial increase in the incidence of recessive diseases on this island.

A small, isolated population also has a qualitative feature. If a population is founded by 20 diploid individuals, no allele can have a frequency less than 1/40 (unless it is 0). Although the frequencies will drift somewhat, genes that are rare in large populations will either be absent (most of the time) or relatively common, with a fairly high probability of homozygosity. Not surprisingly, medical geneticists studying rare conditions like to use populations that for geographical, cultural, or religious reasons have started with a small number and have remained isolated. Diseases that are ordinarily very rare are found in sufficiently high frequency for effective comparative study. Each population has its own special kinds of mutant genes, which are different from those of other isolated populations and are in the population solely because they were present in one of the founders.

Many peculiar traits exhibited by isolated populations of genes can be traced back to the founding population, usually to a single person. One Pacific island has a high frequency of a recessive total color blindness (achromatopsia), characterized by the inability to see in bright sunlight because of excessive glare. The allele is traced back to

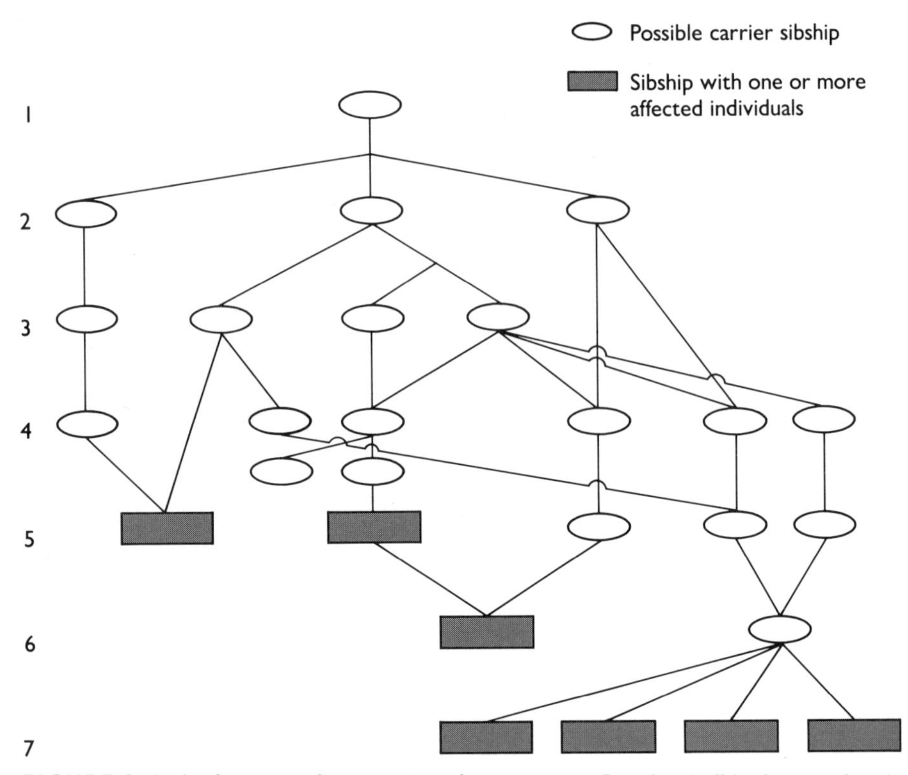

FIGURE 2-6. A schematic pedigree tracing achromatopsia on Pingelap Atoll back to one founder, Miwahuele. Only the sibships, not the individual members of the sibships, are shown. (Data from I. Hussels and N. Morton. 1972. *Amer. J. Hum. Genet.* 24:304.)

an individual who was one of the few people to survive a typhoon several generations in the past (see Figure 2-6).

As another example one of the Caribbean islands has a high incidence of deficiency of apolipoprotein C-II, which is required to activate liproprotein lipase. Homozygous recessives have several metabolic problems, of which pancreatitis is the most critical. The large number of affected persons is caused both by the high level of consanguinity on the island and the presence of the rare gene in one of the founders.

Equilibrium Between Random Drift and Mutation. Equations like 2-7 and 2-8 imply that if we wait long enough, the population will become homozygous. For example, if $N = 100$, then $1 - 1/(2N + 1)$ is 0.995. After 200 generations the heterozygosity will be about 37 percent of its original amount, and after 400 generations about 14 percent. Even a large population, after a very long time, should eventually become nearly homozygous.

Of course, populations do not become entirely homozygous. New mutations continually occur, adding to the population heterozygosity. To modify Equation 2-6 to include mutation, we assume that each new mutant allele does not preexist in the population. This assumption is reasonable if we consider that the coding region of a typical gene is 1000 nucleotides or more in length and that three possible alterations can occur at each site. Therefore at least 3000 single-step mutations are possible, with an enormously larger number if we take into account changes involving more than one nucleotide. Thus any new mutants are unlikely to be duplicates of alleles already in the population, indicating that the model is a good approximation to reality.

We modify Equation 2-5 as follows:

$$F_t = \left[\frac{1}{2N_e} + \left(1 - \frac{1}{2N_e}\right)F_{t-1}\right](1 - \mu)^2, \tag{2-10}$$

where μ is the mutation rate per locus per generation. The extra term $(1 - \mu)^2$ is the probability that neither of the two identical alleles has mutated during the current generation; if they have mutated, they will no longer be identical. As you see, incorporating mutation into the equation is easy. Of course, the correction for mutation will be important only if we are considering the process over many generations.

Eventually this process will reach an equilibrium in which the loss of alleles by random drift is balanced by new mutations. We can specify such an equilibrium by setting $F_t = F_{t-1} = \hat{F}$ and rearranging the equation.

$$\hat{F} = \frac{(1 - \mu)^2}{2N_e - (2N_e - 1)(1 - \mu)^2}. \tag{2-11}$$

Since μ is ordinarily 10^{-5} and often less, we can ignore μ^2, leading to

$$\hat{F} = \frac{1 - 2\mu}{4N_e\mu - 2\mu + 1} \approx \frac{1}{4N_e\mu + 1}. \tag{2-12}$$

A circumflex indicates an equilibrium value.

This analysis holds a pleasant surprise. We have gotten a bonus. If we assume, as we did, that every mutant allele does not preexist in the population, then all alleles that are alike must be descended from a single mutant. Therefore identity by descent and identity in state are synonymous. The bonus is that F now measures not only identity by descent and $1 - F$ *relative* heterozygosity, but F and $1 - F$ also measure *absolute* homozygosity and heterozygosity.

From now on I shall distinguish between absolute and relative measures by using F and *Het* when only relative values are considered and G and H for absolute homozygosity and heterozygosity. Then we can obtain $\hat{H} = 1 - \hat{G}$ by subtracting the right-hand side of Equation 2-12 from 1, giving

$$\hat{H} \approx \frac{4N_e\mu}{4N_e\mu + 1}. \qquad \text{2-13}$$

Figure 2-7 shows a plot of $\hat{H}$ against $4N_e\mu$. The pivotal quantity is $4N_e\mu$. If it approaches 0, individuals are homozygous at most unselected loci and the population is mainly monomorphic. If $4N_e\mu$ is large, say 10 or more, the population is almost entirely heterozygous and many alleles are segregating at each locus.

Suppose that the effective number is half the number of adults, a reasonable value for many populations. Then one new mutant per generation implies that $2N\mu = 1$, or $4N_e\mu = 1$. Thus if the number of mutations at this locus per generation is more than one, the population will be mainly heterozygous; if less than one, it will be mainly homozygous.

The preceding discussion depends on the assumption that the alleles are selectively neutral and that each mutant is new (the "infinite allele model"). From Table 1-9 we saw that the average heterozygosity for isozyme loci in vertebrates is about 0.05 and perhaps twice this amount in many invertebrates. This level of heterozygosity seems to be roughly the same in abundant and rare species (although invertebrates, particularly insects, generally have larger populations than do vertebrates). That no populations approach a heterozygosity anywhere near 1 suggests that the neutral model is not strictly correct. We shall return to a discussion of this topic later in Chapter 7.

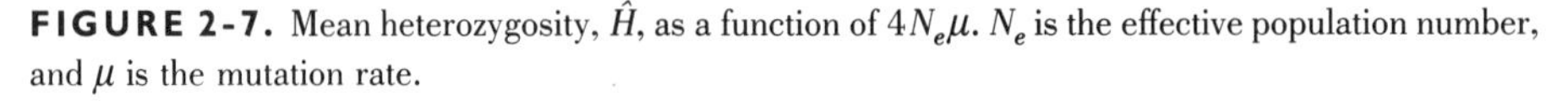

FIGURE 2-7. Mean heterozygosity, $\hat{H}$, as a function of $4N_e\mu$. N_e is the effective population number, and μ is the mutation rate.

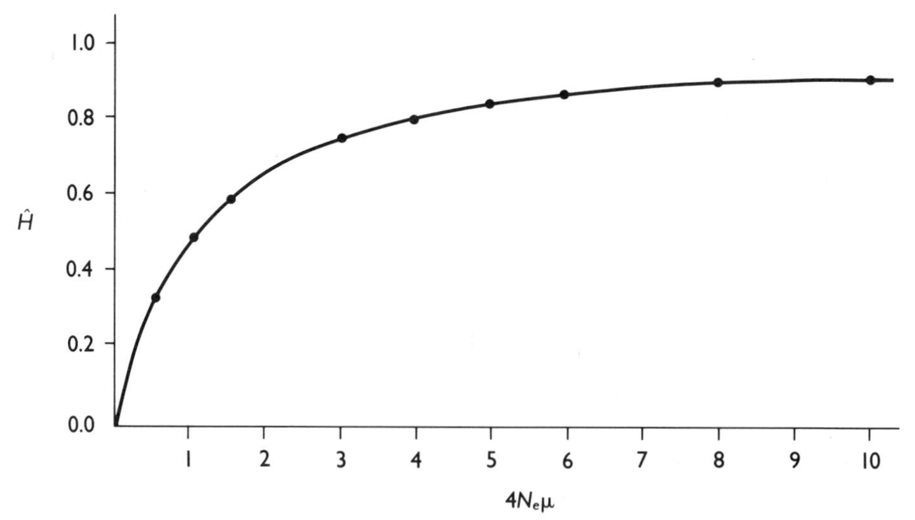

The Rate of Neutral Evolution. As just discussed, $\hat{H} = 1 - \hat{G}$ tells us how much heterozygosity is generated by neutral mutation when the occurrence of new mutants is balanced by their loss from random drift. What happens to an individual mutant? Most of the time it is lost within the first few generations, simply by chance. But occasionally it may drift to a high frequency. Still less often, a mutant may drift to such a high frequency that it replaces the existing allele. How often does such **fixation** happen?

If we consider a *very* long time period, eventually all the genes at a locus will be descended from a single allele. If this outcome seems strange to you, look at it this way. Suppose we start with a pool of n alleles, all different. In the next generation some alleles will have no descendants and others will have more than one, even if the average number of descendants per allele is one. The alleles that left no descendants are permanently lost from the pool. If we continue this process for many generations, all but one allele will disappear, and that one allele will make up the whole population.

If μ is the neutral mutation rate per generation at the locus under consideration, then $2N\mu$ new mutants are produced each generation in a diploid population of size N. Of the $2N$ genes in the population, one gene, or $1/2N$ of the total, will ultimately be fixed. The probability that the fixed gene is a mutant is $2N\mu(1/2N)$, or simply μ.

This seemingly trivial formula answers our question about the frequency of fixation. The rate at which neutral mutants are fixed in the population is simply the mutation rate per generation. Perhaps you are surprised that the rate of fixation in the whole population of genes is the same as the rate of mutation in a single gene. Even more surprising is that the population number does not affect the result. The reason is that although random drift is greater in a small population, the number of new mutants is less, and the two effects cancel.

Figure 2-8 illustrates this process. The figure shows that we must observe the process over a time period that is long relative to the time required for a single fixation to occur. To understand this process, we need to know how long it takes for a surviving mutant to sweep through the population. The derivation of the value is too complex for this text, but the answer is about $4N_e$ (see C&K, pages 430 ff).

Figure 2-8 considers only neutral mutations. Harmful mutations disappear in the first few generations and are not included in the figure. We shall consider favorable mutations in Chapter 7.

The lines of ancestry between humans and fish separated about 400 million years ago. Therefore independent evolution has taken place for 800 million years. The mean number of differences per amino acid in alpha hemoglobin between human and carp is 0.66. Therefore the evolution rate is $0.66/(8 \times 10^8)$, or 0.8×10^{-9}. This value represents about one amino acid change, or one substitution, per codon per billion years.

Of course, we do not know the mutation rate throughout this time period, but according to our fragmentary knowledge of human hemoglobin mutation rates, we can judge the rate of fixation to be considerably less (perhaps $1/10$ as high as the mutation

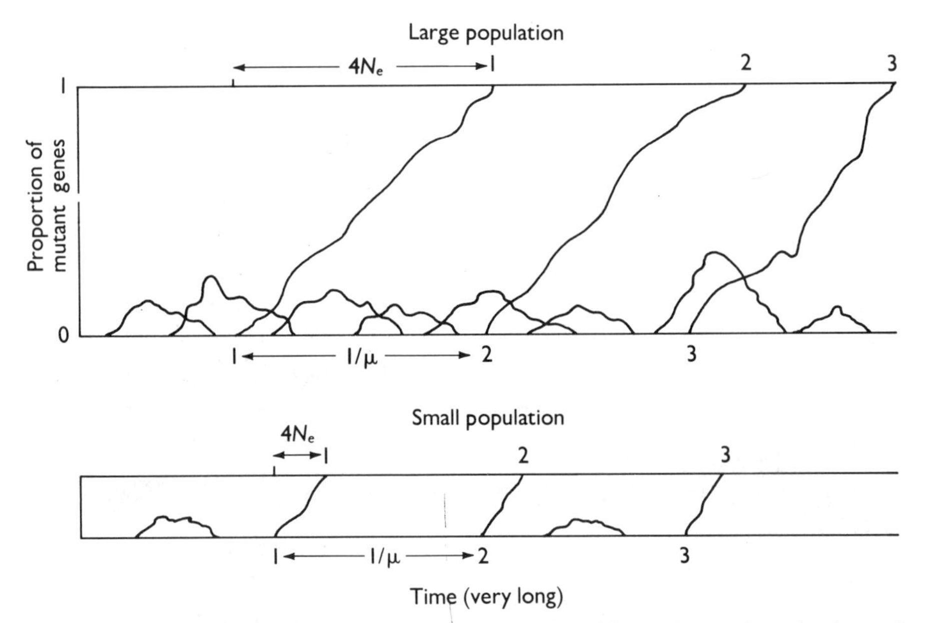

FIGURE 2-8. The fixation of neutral mutations. Comparison of the two figures shows that the number of substitutions is independent of the population size, although the time required for an individual fixation is proportional to the effective population size. The ratio of fixed to lost mutants is grossly exaggerated.

rate). If 1/10 of the mutants were neutral and the rest harmful and quickly eliminated, this would account for the observed rate of amino acid changes.

In the late 1960s several people, most emphatically Motoo Kimura, suggested that most molecular changes in evolution are essentially neutral and accumulate by mutation and random fixation. This hypothesis has generated a great deal of debate, and the evidence is still equivocal. Nevertheless, support for the theory has steadily increased since its first appearance. We shall return to this topic in Chapter 7. Kimura's recent book (1983) on this subject is listed among the references in the back of this book.

2-5 ASSORTATIVE MATING

At the beginning of this chapter, I mentioned that assortative mating had some of the same consequences as inbreeding. Once again, assortative mating is the mating of individuals with similar phenotypes; inbreeding is the mating of individuals with similar genotypes. To the extent that the phenotype is determined by the genotype, we

TABLE 2-1

Genotypes and frequencies for a recessive red-hair gene.

PHENOTYPE	NONRED	NONRED	RED	TOTAL
Genotype	AA	Aa	aa	
Frequency in generation t	P_t	$2Q_t$	R_t	1

Frequency of allele $a = q = Q_t + R_t$

might expect assortative mating to increase homozygosity. It does, but in most situations the effect is very slight.

Let us consider the role of assortative mating in increasing the incidence of recessive traits. A convenient example is red hair in humans, which is usually recessive. Some evidence suggests that red-haired people tend to marry each other. Assume that a fraction, r, of red-haired people mate among themselves, while the remaining $1 - r$ mate randomly with the rest of the population. Table 2-1 shows the genotypes and their frequencies.

The subscript t denotes time in generations, as usual. I have not subscripted q because the allele frequency does not change. Pure assortative mating, like inbreeding, does not systematically change the gene frequencies, but the proportions of the different genotypes will change as the population becomes more homozygous. (Sometimes when strong assortative, or especially disassortative, mating occurs, some genotypes cannot find a mate; hence some genes decrease or increase in frequency. I prefer to regard this situation as a mixture of assortative mating and selection. Assortative mating, like inbreeding, reassorts genotypes but does not change allele frequencies.)

Consider the population to be divided into two groups: (1) those redheads who mate assortatively and (2) the rest of the population. The proportions will be rR_t for group 1 and $1 - rR_t$ for group 2. Among the assortatively mating group, all the children will be aa. In the rest of the population, the frequency of allele a is $[Q_t + R_t(1 - r)]/[P_t + 2Q_t + R_t(1 - r)]$, which reduces to $(q - rR_t)/(1 - rR_t)$. Therefore the proportion of aa genotypes in the next generation will be

$$R_{t+1} = rR_t + \left(\frac{q - rR_t}{1 - rR_t}\right)^2 (1 - rR_t). \qquad \text{2-14}$$

Table 2-2 and Figure 2-9 show the change of frequency of homozygous recessives for several values of the allele frequency, q, and the degree of assortative mating, r. The effect of assortative mating is clearly small unless the degree of assortment is quite

TABLE 2-2

Frequencies of recessive homozygotes in successive generations of assortative mating; *q* is the allele frequency, *t* the time in generations, and *r* is the degree of assortative mating.

	q	0.05	0.10	0.10	0.10
t	*r*	0.25	0.25	0.50	0.75
0		0.0025	0.0100	0.0100	0.0010
1		0.0031	0.0120	0.0141	0.0161
2		0.0032	0.0124	0.0141	0.0199
3		0.0032	0.0125	0.0164	0.0223
4		0.0032	0.0125	0.0167	0.0238
Limit		0.0032	0.0125	0.0169	0.0263

extreme. Probably $r = 0.25$ is too high for red hair; so the increase in frequency is quite slight. Notice that the equilibrium value is attained in a few generations unless the degree of assortment is large. Notice also that the population never becomes completely homozygous unless $r = 1$. This characteristic contrasts with most forms of inbreeding, which eventually lead to complete homozygosity (in the absence of mutation).

To obtain the equilibrium value, we can set $R_{t+1} = R_t = \hat{R}$, yielding the quadratic equation

$$r\hat{R}^2 - (1 - r + 2qr)\hat{R} + q^2 = 0, \qquad \text{2-15}$$

which can be solved in the standard way to give

$$\hat{R} = \frac{1 - r + 2qr - \sqrt{(1 - r + 2qr)^2 - 4rq^2}}{2r}. \qquad \text{2-16}$$

A human trait for which assortative mating is rather high is deafness. Also, a substantial fraction of deafness is caused by recessive alleles. Thus we might expect that the incidence of deafness would be substantially increased by assortative mating. The increase is slight, however. The reason is that although recessive deafness is common, there are many different loci at which homozygosity for a recessive allele leads to deafness. The normal alleles at different loci are complementary, so that children of two deaf parents often have normal hearing because the parents were homozygous at different loci. At least 30 loci are estimated to contain recessive alleles for deafness (Chung, Robison, and Morton, 1959). This large number greatly dilutes the effectiveness of assortative mating in increasing the incidence of deafness.

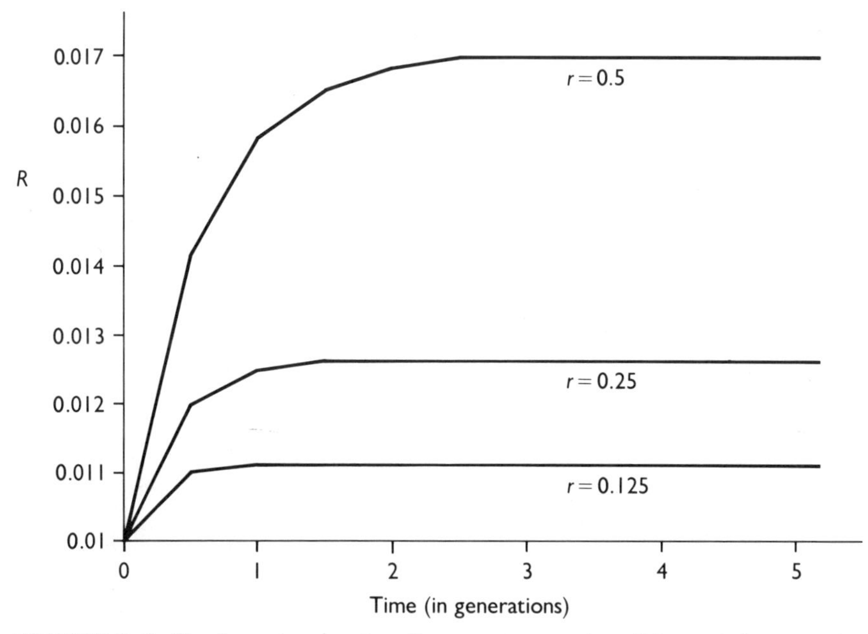

FIGURE 2-9. The change in proportion of homozygous recessives, R, in successive generations with assortative mating of degree r. The allele frequency, which does not change, is 0.1.

In general, we conclude that assortative mating does not significantly increase the incidence of recessive traits. Assortative mating has a much greater effect on multifactorial traits, but the effect is not to increase homozygosity appreciably but rather to increase the population variability. This effect is very important to animal and plant breeders, human geneticists, and students of evolution. We shall return to this topic in Chapter 5, after considering multifactorial inheritance.

QUESTIONS AND PROBLEMS

1. The incidence of PKU, an autosomal recessive form of mental retardation, is 1/10,000 when the parents are unrelated. What is the risk of an affected child if the parents are J and K in Figure 2-10a?
2. (a) What is the inbreeding coefficient of individual I in Figure 2-10b? (b) What is the inbreeding coefficient for an X-linked locus?
3. What is the inbreeding coefficient in individual I in Figure 2-10c if we are given the additional information that the inbreeding coefficient of K is 1/4?

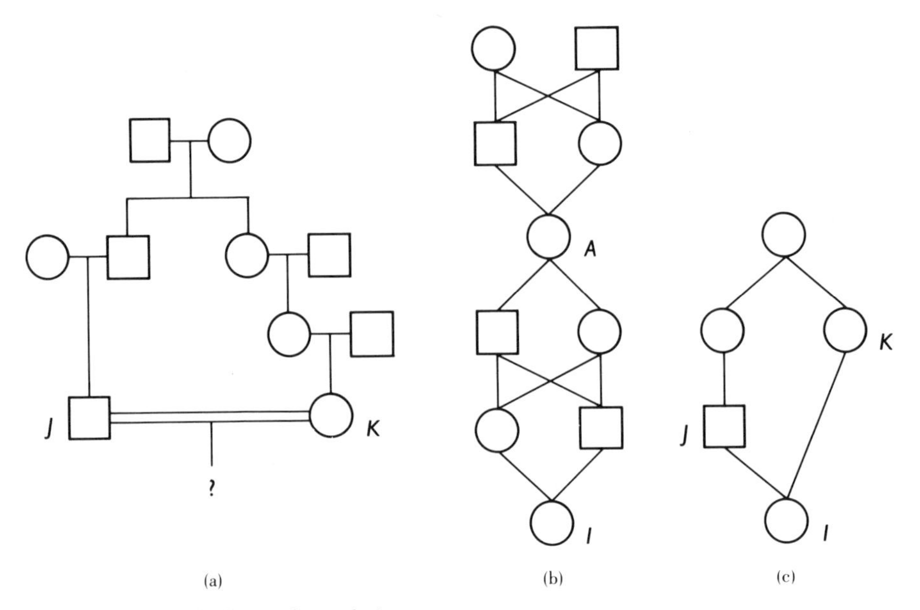

FIGURE 2-10. Pedigrees for analysis.

4. The preadult death rate for children of first cousins is 0.05 higher than for those of unrelated parents. What is the average number of lethal equivalents per person?
5. If the number of lethal equivalents per gamete is 1.5, would you expect self-fertilization to become common?
6. An anthropologist found that, in a population with a complicated system of kin marriages, 17 of 800 people are homozygous for isozyme allele A_S. Later the marriage system broke down and mating became random. The frequency of these homozygotes then dropped to 0.01. What was the average inbreeding coefficient under the old system?
7. Which are most closely related for X-linked loci: brothers, sisters, or brother and sister?
8. If the effective population number for nuclear genes is N, what is it for mitochondrial genes?
9. Would you expect to find a lower incidence of rare recessive diseases in the United States today than 100 years ago? Why?
10. What is the kinship coefficient for a Y-linked locus in two brothers?
11. Invent an algorithm for computing inbreeding coefficients that, instead of introducing the term $(1 + F_A)$ for common ancestors, permits going through the same individual in a path more than once.

12. (a) Give several reasons why the effective population number is usually less than the census number. (b) Under what circumstances is the effective population number larger than the actual number?

13. What is the effective population number of a population that repeatedly cycles through sizes 10, 50, 250, and 1250?

14. Male honeybees are haploids arising from unfertilized eggs, whereas the queen and sterile workers are biparental diploids. Assume that the workers in a colony are the progeny of a single queen and male. (a) What is the kinship of workers? (b) What is the kinship of a worker and the queen? (c) If the queen is the mother of the male, what is the inbreeding coefficient of a worker?

15. *A*, *B*, and *C* are highly inbred strains of mice ($F = 1$). An *AB* hybrid is mated with a mouse from strain *C* and the progeny are mated with each other. What is the inbreeding coefficient of *their* offspring?

16. Is the similarity in the amount of heterozygosity in different-sized populations consistent with neutralism?

17. Is the constancy of the molecular evolution rate per year consistent with neutralism?

18. (a) If the incidence of a rare recessive disease is *I*, what is the approximate incidence in first cousins of affected persons? (b) If the incidence in cousins is less than the expected number, what hypothesis would you suggest?

19. What is the effective number of a population of one male and a large number of females?

20. Cystic fibrosis and alkaptonuria are both caused by recessive alleles. The proportion of alkaptonuric children whose parents are related is much higher than that for cystic fibrosis. Why?

21. Suppose that because of hunting pressure, the number of male deer is reduced to two, whereas the female number remains large. What is the approximate effective population number?

22. Four alleles at a locus have frequencies p_1, p_2, p_3, and p_4. What proportion is expected to be heterozygous in a population with inbreeding coefficient *F*?

23. The formula $H = 4N_e\mu/(4N_e\mu + 1)$ measures absolute heterozygosity, whereas the formula $H_t = H_0(1 - 1/2N_e)^t$ measures only relative heterozygosity. What assumption, other than selective neutrality, is required to make the first formula correct?

24. An individual with two alleles that are identical by descent is sometimes said to be autozygous; without such alleles the individual is allozygous. (a) When is an individual homozygous but not autozygous? (b) When is an individual autozygous but not homozygous?

25. Based on Japanese data, offspring weight decreases by about 1 pound when the parents are first cousins. A shipwrecked population of 20 (10 of each sex) mates at random for five generations with no change in population size. How much would weight be reduced at the end of this time?

26. The infant and childhood death rate is, say, 0.05 for unrelated parents, 0.06 when the parents are second cousins, 0.07 for one-and-one-half cousins, and 0.09 for first cousins. (a) Compute the number of lethal equivalents per diploid individual using each degree of relationship. (b) Are the data consistent with a linear inbreeding effect?

27. What are the consequences of completely assortative mating among three genotypes, AA, AA', and $A'A'$?

28. Contrast the long-time effect of mild assortative mating for a recessive trait and mild inbreeding on the frequency of homozygous recessives.

CHAPTER 3

MIGRATION AND POPULATION STRUCTURE

3-1 MIGRATION

Natural populations spread over a large area are hardly ever genetically homogeneous. In many species the population is made up of local subpopulations, with occasional migrants between them. Typically, mating is approximately random within each subpopulation. How much the subpopulations differ from each other depends on the amount of migration. If there is extensive migration, the whole population approximates a single panmictic unit; if there is little migration, the subpopulations may be quite different. Even if the population is distributed continuously over an area, there may be local differentiation if the population is large relative to the distance that any individual moves during its lifetime. This process is an example of **isolation by distance.**

This chapter is devoted to the effects of isolation in producing genetic differences between subpopulations and the effects of migration in reducing such differences. Local groups, or subpopulations, are often small enough to permit considerable random allele-frequency drift. As local groups become more homozygous, the alleles fixed in the various populations tend to differ, so that the groups drift apart. This is offset if there is migration between the groups. Before considering the opposing effects of random drift and migration, we shall look at a case in which the population is large enough that the effects of random drift can be neglected.

Consider a population in which alleles A and a have frequencies p and q. Each generation, this population exchanges a fraction m of its members with another, much larger population that has frequencies P and Q. Next generation in the smaller population, a proportion $1 - m$, representing the fraction that has not emigrated, will have the same frequency, p, as before. A fraction m, the migrants, will have allele A with frequency P. Altogether, the frequency of A alleles in terms of the frequency of the previous generation will be

$$p_t = (1 - m)p_{t-1} + mP. \qquad \text{3-1}$$

Let us assume that the large population is so big that its allele frequency is not changed by the migration; hence we treat P as a constant and do not subscript it.

We can clarify the meaning of this equation by a now familiar device, subtracting P from both sides and rearranging, to give

$$p_t - P = (1 - m)(p_{t-1} - P). \qquad \text{3-2}$$

As we have done in earlier chapters, we now note that the relationship between p_{t-1} and p_{t-2} is the same as that between p_t and p_{t-1}. Making this substitution and continuing for t generations, we have

$$p_t - P = (1 - m)^t(p_0 - P). \qquad \text{3-3}$$

As t gets large, p gets closer and closer to P. Continual immigration from a large population with frequency P eventually converts the population receiving the migrants to that frequency.

As an example Adams and Ward (1973, *Science* 180:1137) found that the Duffy blood group allele Fy^a has a frequency of 0.422 in Georgia whites and 0.045 in Georgia blacks. The frequency of this allele in West African blacks is essentially 0. The original black population came from Africa to Georgia about 10 generations ago. Letting $t = 10$, $p_0 = 0$, $p_t = 0.045$, and $P = 0.422$, we can solve for m: $(1 - m)^{10} = [(p_t - P)/(p_0 - P)]$. Solving this equation gives $m = 0.011$. Thus the average rate of migration of white genes into this black population has been about 1 percent per generation.

Based on the Duffy allele, the proportion of white ancestry in this black population is $0.045/0.422 = 0.11$. Similar studies in New York, Detroit, and Oakland gave somewhat higher values, in the range of 20 to 25 percent. African genes have also been introduced into the white population, but the effect is smaller because of both the larger number of whites and the tendency to classify persons of mixed background as black.

If continuous migration occurs between two or more populations, clearly each population will eventually have allele frequencies that are the weighted average of the frequencies of the initial populations. The rate of approach to this value depends, of course, on m.

Over longer time periods other processes become influential. If the subpopulations are small, there is loss of alleles by random drift, counterbalanced by the introduction of new alleles by mutation or migration. We shall first consider migration.

3-2 WRIGHT'S ISLAND MODEL

We shall now consider the combined effects of migration and random drift. I shall use the same procedures as were used in deriving Equations 2-12 and 2-13. Assume that the entire population is very large (effectively infinite) and that the subpopulation, or group, of interest has effective number N_e. Each generation, this group exchanges a fraction m of its members for migrants drawn at random from the rest of the population. You can think of this process as an exchange of a fraction, m, of the group's genes. This approach is actually closer to the mathematical model than an exchange of individuals, since the model assumes that incoming alleles are independent. This assumption is not strictly true, for the two alleles in an individual come from the same group and are therefore correlated. However, it can be shown that unless m is quite large the independence assumption leads to a very good approximation. I am also assuming that the number of alleles in the entire population is great enough to ensure that a migrant always brings in alleles not preexisting in the subpopulation.

If we ignore migration we have Equation 2-5. I am using G instead of F to emphasize that we are discussing absolute homozygosity. G_t is the probability that two randomly chosen alleles in a subpopulation in generation t are identical. If one of the alleles is a migrant, they are not identical. The probability that neither of the two chosen alleles is a migrant is $(1 - m)^2$. Including migration in Equation 2-5, we have

$$G_t = \left[\frac{1}{2N_e} + \left(1 - \frac{1}{2N_e}\right)G_{t-1}\right](1 - m)^2. \qquad \text{3-4}$$

I have neglected the remote possibility that the two chosen alleles are both new migrants that happen to be identical. The formulae for mutation versus random drift and for migration versus random drift are the same, although the meanings are different.

We can find the equilibrium value by equating G_t to G_{t-1}, which leads to

$$\hat{G} = \frac{(1-m)^2}{2N_e - (2N_e - 1)(1-m)^2}. \tag{3-5}$$

If m is small, the formulae for homozygosity and heterozygosity become

$$\hat{G} \approx \frac{1}{4N_e m + 1} \quad \text{and} \quad \hat{H} \approx \frac{4N_e m}{4N_e m + 1}, \tag{3-6}$$

which are analogous to Equations 2-12 and 2-13. The key component is $4N_e m$. If this quantity is 4, then $\hat{H} = 4/5$, meaning that the population is mostly heterozygous. If this quantity is $1/4$, then $\hat{H} = 1/5$, and the population is mostly homozygous. If $4N_e m = 1$, the heterozygosity is $1/2$. If $N_e = N/2$, then when $4N_e m = 1$, the absolute number of migrant genes, $2Nm$, is 1. The population structure pivots around one migrant gene per generation. If the number is larger than 1, the subpopulation tends to be heterozygous; if less than 1, the subpopulation tends to be homozygous.

The infinite allele model is not always realistic. Let us modify it to consider a fixed number, k, of alleles. For convenience assume that in the entire infinite population the alleles are equally frequent. Equation 3-4 then becomes

$$G_t = \left[\frac{1}{2N_e} + \left(1 - \frac{1}{2N_e}\right)G_{t-1}\right](1-m)^2 + \frac{2m(1-m)}{k}. \tag{3-7}$$

The rightmost expression is the probability that one of the two chosen alleles is a migrant and the other is not, $2m(1-m)$, multiplied by the probability that these alleles are identical, which is $1/k$ since all incoming migrant alleles are equally frequent. I am ignoring the possibility that both chosen alleles are new migrants.

Equating G_t and G_{t-1} and assuming that m is small, we have the approximate equilibrium formulae

$$\hat{G} \approx \frac{1 + (4N_e m/k)}{1 + 4N_e m} \quad \text{and} \quad \hat{H} \approx \frac{4N_e m[(k-1)/k]}{1 + 4N_e m}. \tag{3-8}$$

As k becomes large, the values of $\hat{G}$ and $\hat{H}$ in Equation 3-8 approach those in Equation 3-6. When there are only two alleles, $\hat{H} = 2N_e m/(4N_e m + 1)$.

It is of interest to see what the actual distribution of allele frequencies looks like. This was first worked out by Sewall Wright for the two-allele case and is shown in Figure 3-1. The derivation is beyond the scope of this book, but see C&K, pages 434 ff.

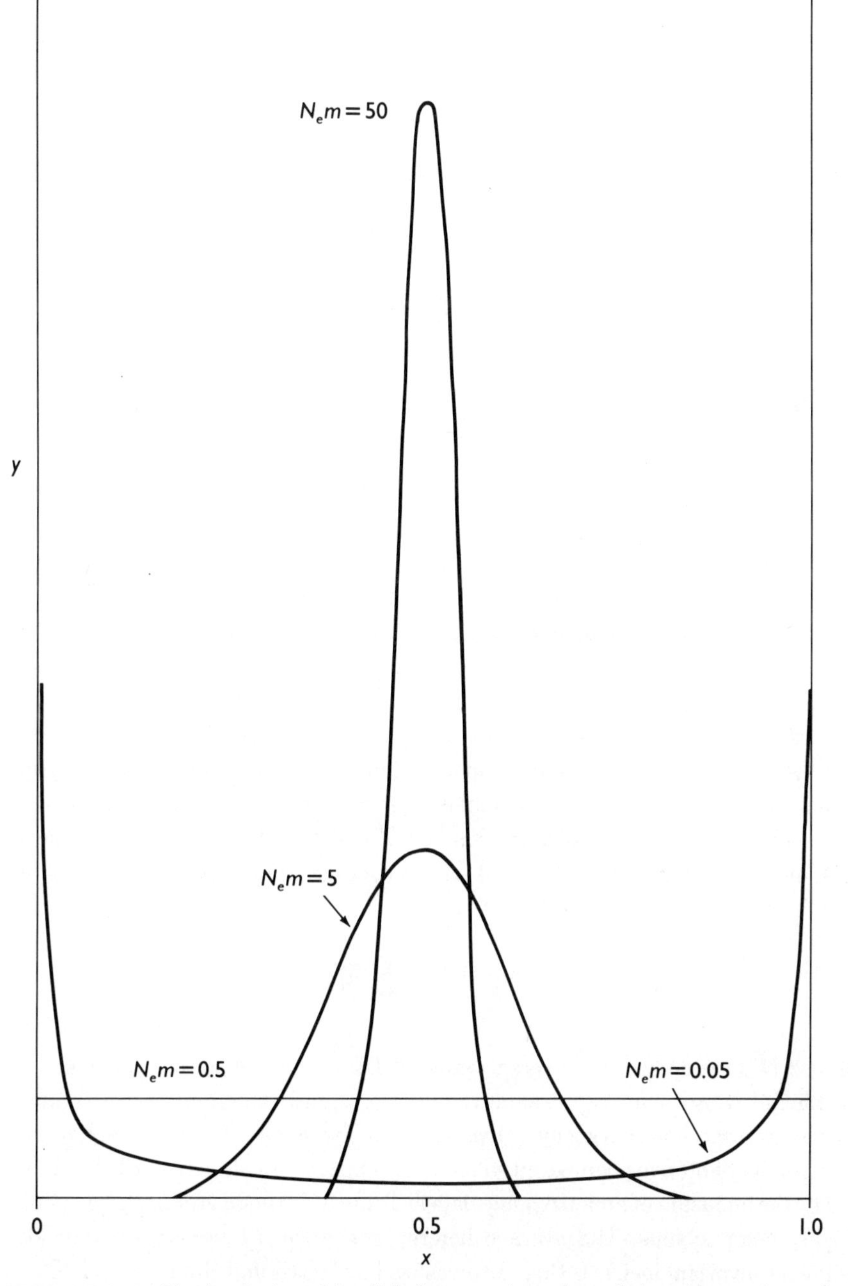

FIGURE 3-1. The distribution of the allele frequency in a subpopulation with mutation and random drift. The allele frequency in the entire population is 1/2. The abscissa is the allele frequency in a subpopulation; the ordinate is the probability density of this frequency; the area under the curves is 1; and m is the proportion of genes in the subpopulation exchanged each generation by migrants.

Wright's work in this area was pioneering, but his model is unsatisfactory in several ways. He assumed that the whole population was infinite and that all of its alleles occurred with equal frequency. Furthermore the model does not include mutation.

I shall soon introduce a less restrictive model, but first we need to digress briefly into the measurement of diversity among populations.

3-3 DIVERSITY AMONG POPULATIONS

As mentioned in Chapter 1, we can measure genetic diversity by calculating the amount of actual or potential heterozygosity. In particular,

$$H = 1 - \sum p_i^2 . \tag{3-9}$$

H is the probability that two randomly chosen alleles are different. If the population is in Hardy-Weinberg proportions, H is a measure of actual heterozygosity. We can use H more generally, however, since it measures genetic diversity whether or not the population is randomly mating; in fact, the population does not even have to be diploid. For example, haploid *E. coli* is about twice as genetically diverse ("heterozygous") as Drosophila.

We can measure the level of subdivision in a large population by comparing the diversity of alleles within a subpopulation to that for the entire population. Let H_S and H_T stand for the diversity in a subpopulation and in the total population. We can measure both these quantities by Equation 3-9. If we let p_{is} stand for the frequency of allele A_i in subpopulation s and $\bar{p}_i$ stand for the average frequency of A_i in the entire population, then H_S and H_T are defined as

$$H_S = 1 - \sum_i p_{is}^2 \quad \text{and} \quad H_T = 1 - \sum_i \bar{p}_i^2 . \tag{3-10}$$

I shall use $\bar{H}_S$ to represent the average value of H_S for several subpopulations.

Table 3-1 shows some representative diversity values. The $\bar{H}_S$ values are equivalent to the average heterozygosity values given in Table 1-9. The generally greater allelic diversity of humans compared with that of other mammals given in Table 1-9 is caused by the inclusion of nonenzymatic data in Table 3-1. Blood groups, for example, are highly heterozygous. The average heterozygosity for 71 human isozyme loci, including 51 invariant loci, is 0.067, as measured by Harris and Hopkinson (1982, *J. Human Genet.* 36:9).

We now have $\bar{H}_S$, which measures the average heterozygosity of the subpopulations, and H_T, which measures the heterozygosity of the total population if it were converted into a single randomly mating unit. We now need a measure of the amount of differentiation among the subpopulations.

The coefficient that I shall employ to measure the extent of differentiation is G_{ST}, first introduced by Masatoshi Nei (see Nei, 1975, for a discussion). It is defined as

$$G_{ST} = \frac{H_T - \overline{H}_S}{H_T}. \qquad \text{3-11}$$

Table 3-1 includes some representative values of G_{ST}. The data on human races are striking: only 7 percent of the diversity is between races; 93 percent of the total variability in the species is found within races. Races differ very little in biochemical polymorphisms. The adaptive significance of such traits is largely unknown. The results would be quite different if the data were based on such traits as skin color, which presumably represents adaptations to different climates.

Another example of differentiation among races is lactose tolerance. Groups that have not traditionally used dairy products, such as Chinese, Tai, Ibo, and West Africans, generally have a low tolerance for lactose. Groups that have used milk for thousands of years, such as European populations, have a high tolerance. Whether measures of group differentiation are high or low depends on the loci chosen for study.

The Yanomama are a group of Brazilian Indians that live in small villages. As the data in Table 3-1 show, the heterozygosity $\overline{H}_S$ is quite low, presumably the result of the small effective number of the villages. Yet the diversity in the entire group of villages is not much larger than that of a single village. The entire population is likely to be descended from a small number of founders some time in the past. If a population goes through a size bottleneck, it may require a very long time to reach its original heterozygosity. How long? That and related questions are discussed in the next section.

TABLE 3-1

Analysis of genetic diversity within and among groups for several species.

SPECIES	NUMBER OF POPULATIONS	NUMBER OF LOCI	H_T	$\overline{H}_S$	G_{ST}
Human—major races	3	35	0.130	0.121	0.070
Yanomama Indian villages	37	15	0.039	0.036	0.069
House mouse	4	40	0.097	0.086	0.119
Kangaroo rat	9	18	0.037	0.012	0.674
Drosophila equinoxialis	5	27	0.201	0.179	0.109
Horseshoe crab	4	25	0.066	0.061	0.072
Club moss	4	13	0.071	0.051	0.284

Data from Nei. 1975.

3-4
THE FINITE ISLAND MODEL

If the total population is finite and mutation is absent, the entire population will eventually become completely homozygous, regardless of structure. If there is migration, a single allele will be fixed throughout; if there is no migration, each group will have its characteristic allele. Yet, because of mutation, this does not happen. We now look at the simultaneous operation of migration, mutation, and random drift.

I shall first derive the necessary relationships using the infinite allele model, as in Section 3-2, and then modify the results to include a finite number of alleles. Let G_S be the probability that two alleles in the same subpopulation are identical (in the literature this quantity is often symbolized by f_0 or F_0). Assume there are n subpopulations all of effective size N_e, so that the total population has effective size nN_e. Let G_D be the probability of identity of two alleles drawn from different subpopulations. You may find it mnemonically useful to think of S and D as standing for "same" and "different." I shall first write the recurrence equations and then explain them.

$$G_{S,t} = \{a[c + (1 - c)G_{S,t-1}] + (1 - a)G_{D,t-1}\}(1 - \mu)^2. \tag{3-12}$$

$$G_{D,t} = \{b[c + (1 - c)G_{S,t-1}] + (1 - b)G_{D,t-1}\}(1 - \mu)^2. \tag{3-13}$$

$$a = (1 - m)^2 + \frac{m^2}{n - 1} \qquad b = \frac{1 - a}{n - 1} \qquad c = \frac{1}{2N_e}. \tag{3-14}$$

Figure 3-2 shows the relevant migration patterns. We let

a = the probability that two individuals drawn from the same population were in the same population a generation earlier.

b = the probability that two individuals drawn from different populations were in the same population a generation earlier.

$1 - a$ and $1 - b$ are the respective probabilities that the two individuals were not in the same subpopulation in the previous generation.

In Figure 3-2 row a shows the two events that together have probability a: (1) the two chosen alleles were in the same group the previous generation and neither has migrated or (2) the two chosen alleles were in the same group the previous generation and both have migrated to the same group. The two probabilities are $(1 - m)^2$ for event 1 and $m^2/(n - 1)$ for event 2; m^2 is the probability that both alleles are in migrants, and $1/(n - 1)$ is the probability that both alleles migrate to the same subpopulation. Similarly row b has two components: (1) the two chosen alleles were in the same group the previous generation and one has migrated or (2) the two chosen alleles were in the same group the previous generation and have migrated to different

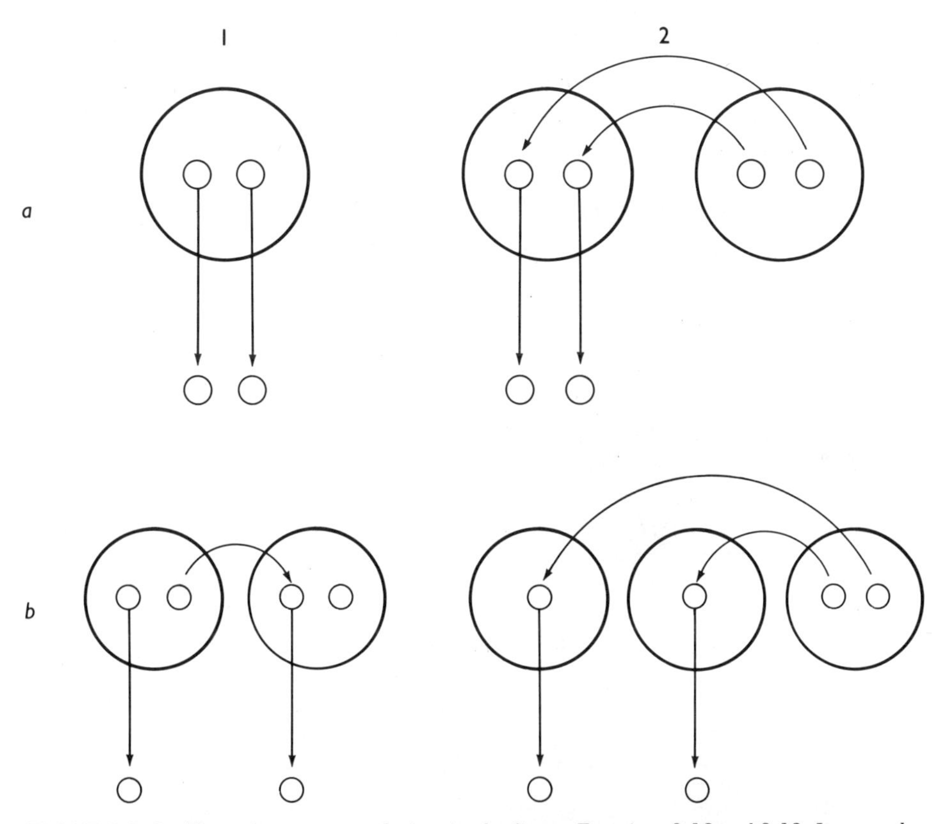

FIGURE 3-2. The various patterns of migration leading to Equations 3-12 and 3-13. In row a the randomly chosen individuals, indicated by the two circles at the bottom, come from the same population. In row b they come from different populations. The curved arrows represent migration in the previous generation.

groups. The two probabilities are $2m(1 - m)/(n - 1)$ for event 1 and $m^2(n - 2)/(n - 1)^2$ for event 2, and b is their sum. It turns out that $b = (1 - a)/(n - 1)$.

Combining these expressions leads to Equations 3-12 and 3-13. The final term $(1 - \mu)^2$ is the probability that neither of the chosen alleles has mutated since the previous generation.

We can begin with any arbitrary starting values for G_S and G_D and derive the values in any later generation by repeated application of these equations. This procedure lets us determine how rapidly the values change, and if we continue long enough, what the equilibrium is. These computations are tedious and best done by computer.

We can also find the equilibrium by equating the values for t and $t - 1$ and solving the two simultaneous equations. We are interested in those groups that have some local differentiation but are not completely isolated. Therefore we want the subpopulations to be small enough for random drift to occur and for migration to be somewhat less than one migrant per generation. Mathematically we assume that $\mu \ll m,\ 1/N_e \ll 1$. Therefore we neglect terms of order μ^2, μm, μ/N_e, m^2, and m/N_e, which greatly simplifies the formulae. We then obtain

$$\hat{G}_S = \frac{m + \mu(n-1)}{m + 4N_e mn\mu + (n-1)\mu} = 1 - \hat{H}_S. \tag{3-15}$$

$$\hat{G}_D = \frac{m}{m + 4N_e mn\mu + (n-1)\mu} = 1 - \hat{H}_D. \tag{3-16}$$

The average diversity in the whole population is the average of n subpopulations, $\hat{H}_S$, and $n(n-1)$ pairs of different populations, $\hat{H}_D$, for a total of n^2. Canceling n in the numerator and denominator leads to

$$\hat{H}_T = \frac{\hat{H}_S + (n-1)\hat{H}_D}{n}. \tag{3-17}$$

We can obtain $\hat{G}_{ST}$ by substituting Equations 3-15 and 3-17 into Equation 3-11, which, if we ignore small quantities, leads to

$$\hat{G}_{ST} = \frac{1}{4N_e m\alpha + 1} \qquad \alpha = \left(\frac{n}{n-1}\right)^2. \tag{3-18}$$

Something remarkable has happened here. The term for mutation rate has canceled out. Furthermore $[n/(n-1)]^2$ is very nearly 1 when the total population is large. The value of $\hat{G}_{ST}$ at equilibrium is determined almost entirely by the absolute effective number of migrants, $N_e m$.

Table 3-2 gives data from population studies of various macaques and baboons. The values of G_{ST} range from 0.02 to 0.11. The mean value is about 0.07, which indicates that the members of a troop are about 93 percent as diverse as members of the entire population. The values are reasonably close to those obtained for some human tribes.

Of course, the data are subject to large errors and should be regarded as only rough estimates. Primates are now being studied extensively, and we can expect more data in the future.

The last column of Table 3-2 gives the **coefficient of relationship,** to which we shall return in Chapter 7. The relationship coefficient is a measure of the fraction of shared genes and is computed as $r = 2G_{ST}/(1 + G_{ST})$. I shall use this coefficient to estimate the amount of competition between groups, by which traits, such as altruistic behavior, which are beneficial to the group but harmful to the individual, might evolve.

Sewall Wright is responsible for the classical development of this subject. He

TABLE 3-2

Coefficient of genetic differentiation, G_{ST}, and coefficient of relationship, r, for several primate species of the genera *Macaca* (macaques) and *Papio* (baboons). Troops are regarded as subpopulations.

SPECIES	NUMBER OF LOCI	NUMBER OF TROOPS	G_{ST}	r
M. fascicularis	7	8	0.062	0.116
M. fuscata	12	33	0.091	0.167
M. sinica	10	12	0.106	0.191
P. anubis	7	12	0.064	0.121
P. cynocephalus	2	10	0.067	0.125
P. hamadryas	8	5	0.023	0.045

Data from K. Aoki and K. Nozawa. 1984. *Primates* 25:171.

analyzed the relationships between correlations, variances, and relative heterozygosity and defined three values:

F_{IT} is the correlation between the value of two alleles in an individual relative to the whole population.

F_{IS} is the correlation between the value of two alleles in an individual relative to its subpopulation.

F_{ST} is the correlation between the value of two random alleles from the same subpopulation relative to the total population.

I have given Wright's formulae because they are widely used in studies of natural populations. Yet with the increasing use of molecular markers, the G and H statistics will likely become more widely applicable. For further reading see Wright (1951). A particularly lucid account of Wright's measures, expressed in terms of relative heterozygosities, is found in Hartl (1981), pages 68 ff.

3-5 SOME DESIRABLE PROPERTIES OF G_{ST}

G_{ST} has several properties that make it particularly useful for assessing the degree of population subdivision. Their derivation is beyond the scope of this book, but I shall summarize some of the conclusions. For the details, see Crow and Aoki (1984).

First, the infinite allele model is not always realistic. If there are k alleles, we can modify Equations 3-12 through 3-16, provided we assume that mutation rates from any allele to any other allele are the same. A striking conclusion emerges: Equation 3-18 is not changed. So G_{ST} is robust with respect to the number of alleles. Furthermore, if the mutation rates are not equal, the relationship is still a satisfactory approximation, assuming that the largest mutation rate is small relative to the migration rates. Since the mutation rate does not enter Equation 3-18, we do not need to know the mutation rates provided they are small relative to m and $1/N$.

Second, we need to consider the rate of approach to equilibrium. G_T and G_S approach equilibrium very slowly, since they depend on the mutation rate and the total population size. Thus hundreds of thousands or millions of generations may pass before G_T and G_S reach values near their equilibrium values. Such equilibria are seldom, if ever, attained; the conditions determining the equilibrium value are likely to change faster than the rate of approach to equilibrium.

If the population goes through a size bottleneck, the earlier heterozygosity may not be restored for a very long time. Several animal species have very low heterozygosity. One example is the cheetah. Its population size must have been reduced to a very small number sometime in the past. Another example is the California sea lion. In this case we know the history. This species was nearly exterminated around the turn of the century. It has now multiplied into a large population, but the reduced heterozygosity still persists.

On the other hand, G_{ST} approaches fairly close to equilibrium values (near enough to be useful practically) in a much shorter time. This time period is determined by m and N. Specifically the time required to go halfway to equilibrium is approximately $2m + 1/2N$, regardless of the number of alleles. Thus many small populations will be fairly close to equilibrium values. Recovery, following a change of conditions, is rapid.

A third useful property of G_{ST} is that we do not need to know much about the total population, only whether or not a large number of subpopulations exists. Provided n is large, α is very close to 1; so the actual number of subpopulations need not be known.

Finally the two parameters N_e and m enter Equation 3-18 only as a product. In many circumstances the absolute number of migrants, $M = Nm$, is better known than m, because of uncertainty about the subpopulation size. Furthermore the ratio of N_e/N may be easier to estimate than either of its components. Since $M(N_e/N) = N_e m$, we can estimate G_{ST} from the absolute number of migrants and the ratio of the effective to the census number. For a more complete discussion of effective population numbers, see C&K, pages 345 ff.

3-6 OTHER MODELS OF POPULATION STRUCTURE

The least realistic feature of the island model is the assumption that immigrants come randomly from the rest of the population. Actually immigrants are more likely to come from subpopulations close by. Another model is the "stepping-stone" model of Kimura. This model assumes that the subpopulations are arrayed in a rectangular grid and that immigrants into a group come from one of the four adjacent groups. This represents an opposite extreme to the island model.

Unfortunately a finite stepping-stone model is mathematically intractable. We have to rely on numerical solutions of special cases. Many such numerical studies have been made. The desirable properties of G_{ST} in the island model appear to hold for the stepping-stone model, in particular the near independence of the mutation rate. Although G_{ST} is dependent on the total population number, it is only weakly so, and we need only a crude estimate of the total population size. Finally, the difference in the effectiveness of migration between the two models is not as great as one might expect. A single migrant in an island model is about as effective as 1.2 to 2.0 migrants from adjacent colonies. The two models are not very different in their effect on G_{ST}. Since most natural populations are probably between these extremes, we can have some confidence in conclusions reached from measurements of G_{ST}.

If the population is dispersed over an area that is much more long than wide, the degree of local differentiation is much greater. For example, we can expect fish in a river or animals along a shore to diverge more per unit distance than animals inhabiting a more compact area.

Finally, although many populations are essentially continuous, each individual is of limited motility. This limitation is particularly true of many plants that cover wide areas. Methods have been developed for studying isolation by distance in a geographically continuous population—again Wright was the pioneer, but his work has been greatly extended in recent years. I shall not discuss these methods here, however. The mathematics is very heavy going, and so far the applications are rather limited. We can expect substantial future advances by the interaction of sophisticated molecular and mathematical techniques.

QUESTIONS AND PROBLEMS

1. Each generation 3 percent of a population is replaced by migrants from a large population. The original frequencies of allele A are 0.1 in the small population and 0.9 in the large one. Assuming that the frequency in the large population does not change, what is the allele frequency in the small population 10 generations later?
2. Show that when m is small, Equation 3-5 reduces to $\hat{G} = 1/(4N_e m + 1)$.
3. In measuring diversity within a population, why is Equation 3-9 used instead of counting the actual proportion of heterozygotes?
4. From the data in Table 1-2, compute H_S for each population, the average value of H_S, H_T, and G_{ST}.
5. If $N_e = N/2$ and N is large, what absolute number of migrants per generation is required to make H equal to half its maximum value?
6. It may be contrary to your intuition that a given number of migrants has the same effect regardless of the population size. How might you explain this in words?
7. Verify Equation 3-18. In particular, note that μ drops out.
8. Compute H for the values in Figure 3-1.

CHAPTER 4

SELECTION

We have considered the concept of allele frequency and have seen how random drift can change frequencies and how isolation can lead to local differentiation of populations. The gene frequency changes have usually been erratic. We shall now consider the systematic changes brought about by selection. Random fluctuations introduce "noise" into the system, but for now I am assuming that the population is large enough to keep such effects small relative to the effects of selection.

The adaptations of organisms to their environments are many and marvelous. The special ways in which desert animals and plants conserve water, the power and agility of carnivores, the speed and alertness of herbivores, the streamlined shape of a fish, the elephant's incredible trunk, a bird's feathers, plants that trap insects, devices for assuring cross-pollination, and the adaptation that impresses us most, the human mind, all attest to the effectiveness and meticulousness of natural selection.

Geneticist J. H. Gerould of Dartmouth College unwittingly performed a beautiful experiment showing how well adapted the normal members of a species are and how maladapted a mutant phenotype is. He reared the green-colored larvae of cabbage butterflies along with some bluish mutants in a greenhouse. To human eyes the mutants look very much like the wild type. One day a pane of glass in the greenhouse broke and some birds got in. Almost unerringly, they picked off the mutants, leaving the wild type. Gerould had been unaware of how protective the normal color was until this single instance showed him most impressively, and distressingly, for he nearly lost his precious mutant strain.

We now want to consider the role of selection in adaptive evolution. We also want to consider how selection changes gene frequencies in populations, how rapidly deleterious genes are eliminated, and why some harmful genes persist. Artificial selection for performance and yield, which depend mainly on multifactorial inheritance, is deferred until the next chapter. In this chapter we consider mainly a single locus and regard generations as discrete and nonoverlapping.

4-1 COMPLETE SELECTION

Complete selection means that one phenotype does not contribute any genes to the next generation, because it fails to survive to the reproductive age or fails to reproduce. A good example occurs in Aberdeen-Angus and Holstein cattle when the breeder culls all red or red-and-white calves from the herd. Those that reproduce are all of the black or black-and-white type. The same result occurs in nature when a lethal or sterilizing genotype is involved.

If the unwanted phenotype is caused by a dominant allele, we can easily see what will happen. If *AA* and *Aa* are eliminated or prevented from reproducing, the remaining individuals are all *aa*. Since all the progeny will be recessive homozygotes like the parents, the unwanted phenotype will no longer be found. Such individuals may occur by mutation, however, which we shall consider later in the chapter.

If the unwanted phenotype is caused by an X-linked recessive allele, *a*, the situation is almost as simple. We first notice that if no *aa* females or *a* males reproduce, no *aa* females will occur next generation. From that time on, the only individuals that show the recessive phenotype will be hemizygous males who received the recessive allele from heterozygous mothers. A heterozygous female may transmit the recessive allele either to a son or to a daughter, with equal probability. If it is transmitted to a son, it is eliminated because affected males do not reproduce. If it is transmitted to a daughter, it is retained. The proportion of heterozygous females is reduced by half each generation, since only half the daughters of heterozygous females are heterozygous. Since affected males come from heterozygous mothers, they too are reduced by half each generation. To summarize:

With complete selection against an X-linked recessive allele, after the first generation of selection the proportion of affected individuals is reduced by half each generation.

Selection against an autosomal recessive is slightly more complicated. I shall again use the gene pool model, which is set forth in Table 4-1.

We calculate the frequency of the recessive alleles that are contributed to the next generation in the same way that we counted genes in Chapter 1. The proportion of a alleles is half the contribution from heterozygotes divided by the contribution of all individuals with the dominant phenotype, or $pq/(1 - q^2) = q/(1 + q)$. The proportion of homozygous recessives next generation with random mating is the square of the allele frequency, or $[q/(1 + q)]^2$.

Now let us designate the recessive allele frequency at the beginning, generation 0, by q_0 and in successive generations by q_1, q_2, and so on. Then

$$q_1 = \frac{q_0}{1 + q_0}$$

$$q_2 = \frac{q_1}{1 + q_1} = \frac{q_0}{1 + 2q_0}.$$

Continuing this pattern for t generations, we have

$$q_t = \frac{q_0}{1 + tq_0}. \qquad 4\text{-}1$$

The frequency of homozygous recessives is

$$q_t^2 = \left(\frac{q_0}{1 + tq_0}\right)^2. \qquad 4\text{-}2$$

Figure 4-1 depicts the results of complete selection.

TABLE 4-1

Complete selection against a recessive allele.

	GENOTYPE				
	AA	*Aa*	*aa*	TOTAL	RECESSIVE ALLELE FREQUENCY
Frequency at the zygote state	p^2	$2pq$	q^2	1	q
Relative contribution to next generation	p^2	$2pq$	0	$1 - q^2$	$\frac{q}{1 + q}$

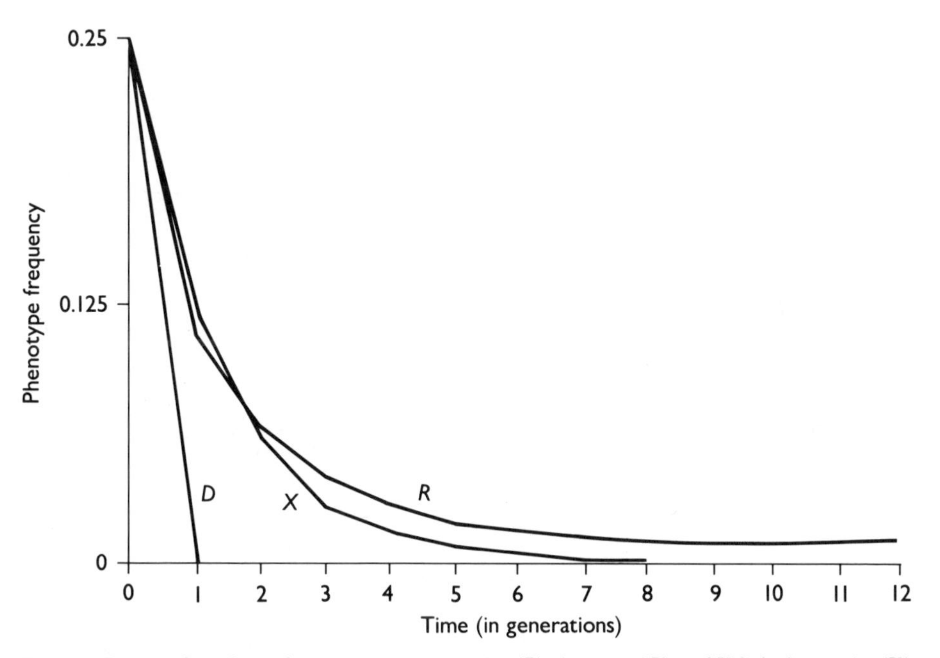

FIGURE 4-1. Complete selection against a recessive (R), dominant (D), and X-linked recessive (X), starting with a phenotype frequency of 0.25. In the X-linked case we assume that there are no homozygous recessive females, as would be the case after one generation of selection.

Now we can easily see why unwanted recessive genes persist at low frequencies for such a long time. As q gets small, the rate of elimination becomes exceedingly slow. Despite breeders having assiduously removed every red-and-white Holstein calf from the breeding population for more than a century, this type still turns up. In Holland, where red-and-white Holsteins are acceptable, this color is quite common. As a numerical illustration, suppose that 1 calf in 400 is red. Thus $q_0 = 1/20$. In 10 generations the allele frequency would be reduced to 1/30 and the red phenotype frequency to 1/900. If the initial allele frequency were 1/100, 10 generations of selection would change the allele frequency to 1/110; the phenotype frequency would have changed from 1/10,000 to 1/12,100, about a 17 percent reduction.

You may have noticed that if we let $q = 1/x$, we can rewrite Equation 4-1 as $1/x_t = 1/(x_0 + t)$. The denominator increases by 1 each generation.

This analysis also shows why serious, even fatal, recessive alleles are not more rapidly eliminated from the population. The frequency of the infant disease galactosemia (inability to utilize lactose) is about 1/50,000. If these infants did not survive and reproduce, as happened before the disease was understood, the incidence next generation would be reduced to 1/50,448, a change of only about 1 percent. This small loss of mutant alleles could well be balanced by new mutations.

In Chapter 2 we considered the deleterious effects of inbreeding caused by the

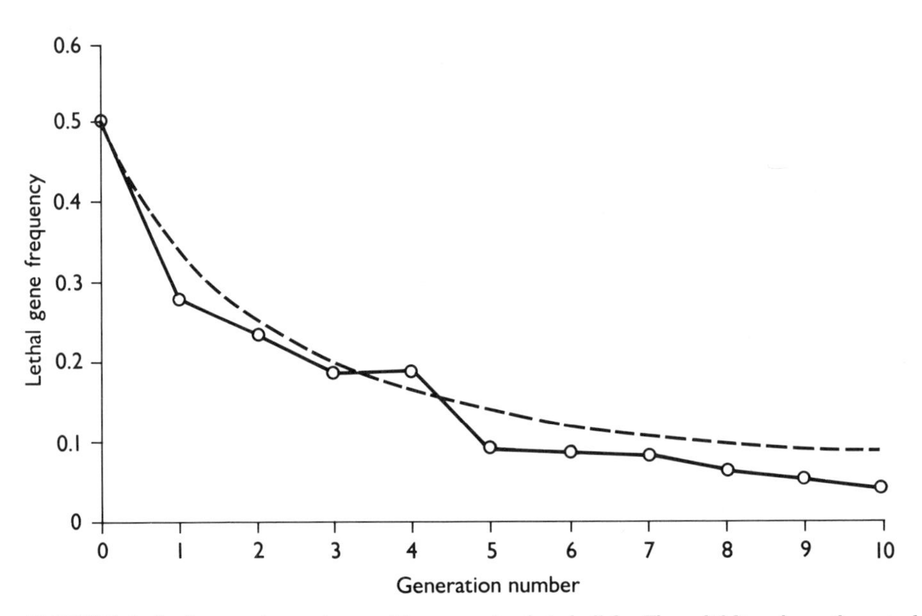

FIGURE 4-2. An actual experiment with a recessive lethal allele. The solid line shows the actual frequencies, whereas the dashed line gives the theoretical expectation. (Data from B. Wallace. 1963. *Amer. Natur.* 97:65.)

exposure through homozygosity of previously hidden recessive alleles. We now see why harmful recessive alleles are so common. Mutants, which are generally harmful, remain in the population a long time if they are recessive but are quickly eliminated if they are dominant.

The Drosophila data in Figure 4-2 show how well the theory predicts actual results. The initial population consisted entirely of heterozygotes; so the initial allele frequency was 1/2. The expected frequencies in successive generations are 1/2, 1/3, 1/4, 1/5, and so on, which you can confirm from Equation 4-1. The agreement is quite good. Since each generation depends only on the previous one, it is not surprising that, after the unexplained drop at generation 5, the frequency continues to remain below the theoretical dotted line. Note that if you compute successive values starting at 1/10 in generation 5 rather than at 1/7, the expectation is very close to the observed points.

4-2 SELECTION OF ARBITRARY INTENSITY

Selection, whether artificial or natural, is usually less intense than in the examples just considered. Instead of one genotype or phenotype not leaving *any* progeny in the next generation, one phenotype may be less effective in survival and reproduction than another. The situation may be further complicated by incomplete dominance.

I shall use **fitness** to mean the combined ability to survive and reproduce. If we start out with two types of zygote, 100 of each type, and the first group leaves 100 descendants in the next generation while the second leaves 95, we say that the *relative* fitness of the second group is 0.95. Again we assume random mating and use the gene pool model. We assume that each parental genotype contributes to the gene pool in proportion to its fitness, and then pairs of genes are chosen at random from this pool.

Of course, a real population does not behave this way. The fertility of an individual may depend on the fertility of its mate; for example, in a strictly monogamous society, if one of the two mates is sterile, both are. Furthermore rather than being constant, the relative fitness of a genotype is likely to depend on what other genotypes are present, the degree of crowding, temperature, and many other factors. In addition, populations in nature usually do not have discrete generations, although some do, such as annual plants and 17-year cicadas. Yet our simple model has the very pleasing property of yielding simple algebraic expressions. It gives answers that are a good approximation to reality, especially when selection is weak, and it gives us a qualitative insight as to how selection changes a population.

General Model of Selection. Table 4-2 shows a general selection model. We assume that the population is mating at random so that the genotypes will be in Hardy-Weinberg ratios at the zygote stage and until there has been any differential mortality. We count each generation at the same stage, as zygotes or before any selective mortality has occurred. Of course, we can develop models for selection when the population is censused at the adult stage, but we shall consider only the simpler model here.

The fitness, w_{11}, is half the mean number of progeny of individuals of genotype A_1A_1; I use 1/2 here because half of each progeny is credited to each parent. Then $\bar{w}$ is half the mean number of progeny per individual in the entire population. If the population is of constant size, $\bar{w} = 1$.

The proportion of A_1 alleles next generation will be the entire contribution of A_1A_1

TABLE 4-2

A general model of selection at a single locus. Each generation is counted at the zygote stage (or before any differential mortality) and random mating is assumed.

GENOTYPE	A_1A_1	A_1A_2	A_2A_2	TOTAL
Frequency	p_1^2	$2p_1p_2$	p_2^2	1
Fitness	w_{11}	w_{12}	w_{22}	
Contribution to gene pool	$p_1^2w_{11}$	$2p_1p_2w_{12}$	$p_2^2w_{22}$	$\bar{w}$

$$\bar{w} = p_1^2w_{11} + 2p_1p_2w_{12} + p_2^2w_{22}$$

genotypes plus half that of A_1A_2, divided by the total contribution of all genotypes, $\bar{w}$ (see the bottom of Table 4-2). Using a prime to stand for the next generation, we have

$$p' = \frac{p_1^2 w_{11} + p_1 p_2 w_{12}}{\bar{w}}. \qquad 4\text{-}3$$

If we have numerical values for the fitnesses and the initial value for p_1, we can crank out the proportions of the three genotypes for all subsequent generations by repeated application of Equation 4-3. The equation gives the allele frequencies, and we use the Hardy-Weinberg principle to get the frequencies of the three genotypes next generation.

We sometimes want to express the change in p_1 each generation. Letting $\Delta p = p' - p$, and using Equation 4-3, we arrive, after some algebraic rearrangement, at

$$\Delta p_1 = p_1 p_2 \frac{(w_{11} - w_{12})p_1 + (w_{12} - w_{22})p_2}{\bar{w}}, \qquad 4\text{-}4$$

where

$$\bar{w} = p_1^2 w_{11} + 2p_1 p_2 w_{12} + p_2^2 w_{22}.$$

A Different Parameterization. Although the model of Table 4-2 is correct and useful, we can obtain formulae with more transparent meanings by changing the symbols. Also, in most genetic applications we are not interested in the total population number but only in the proportions of the different alleles and genotypes. Therefore since we are interested in only relative fitnesses, we can let one of the fitnesses be 1 and measure the others relative to this value.

Throughout this chapter I shall let A stand for the allele that is favored by selection and A' for the alternative allele. To avoid too many subscripts, I shall use p for the frequency of A and q $(= 1 - p)$ for A'. It is also convenient to assign a fitness of 1 to the favored genotype and measure the others relative to this value. Therefore let the relative fitnesses of AA, AA', and $A'A'$ be 1, $1 - hs$, and $1 - s$. If s is 0.05, then the $A'A'$ genotype is 95 percent as fit as AA. The letter h is a measure of heterozygous effect, or dominance. When $h = 0$, AA and AA' have the same phenotype, and thus the A allele is dominant. When $h = 1$, A is recessive. When $h = 1/2$, neither allele is dominant, a situation sometimes called semidominance. Thus h can take any value between 0 and 1; in fact, it can be negative if the heterozygote is more fit than the best homozygote. The model is set forth in Table 4-3.

As before, the proportion of A alleles next generation, p', will include the total contribution of AA genotypes plus half the contribution of AA' genotypes, divided by

TABLE 4-3

An alternative parameterization of the model in Table 4-2.

GENOTYPE	*AA*	*AA'*	*A'A'*	TOTAL
Relative fitness	1	$1 - hs$	$1 - s$	
Frequency at zygote stage	p^2	$2pq$	q^2	1
Relative contribution to the gene pool	p^2	$2pq(1 - hs)$	$q^2(1 - s)$	$\overline{w}$

$$\overline{w} = p^2 + 2pq(1 - hs) + q^2(1 - s) = 1 - sq(q + 2ph)$$

the total contribution of all genotypes, $\overline{w}$. Thus Equation 4-3 becomes

$$p' = \frac{p^2 + pq(1 - hs)}{\overline{w}} = \frac{p(1 - qhs)}{1 - sq(q + 2ph)} \qquad \text{4-5}a$$

or, in the alternative form analogous to Equation 4-4,

$$\Delta p = \frac{spq[q + h(p - q)]}{\overline{w}}. \qquad \text{4-5}b$$

The example in Figure 4-3 shows that these equations work very well in real populations. The data come from a Drosophila population cage. The flies are of three

FIGURE 4-3. The effect of selection on an allele that produces stubble bristles when heterozygous and is lethal when homozygous. The ordinate is the frequency of the *Sb* allele in adult flies. The dashed line is the expectation when there is no selection against heterozygotes; the solid line represents the 12 percent selective disadvantage of heterozygotes. (Data from Y. J. Chung. 1967. *Genetics* 57:957.)

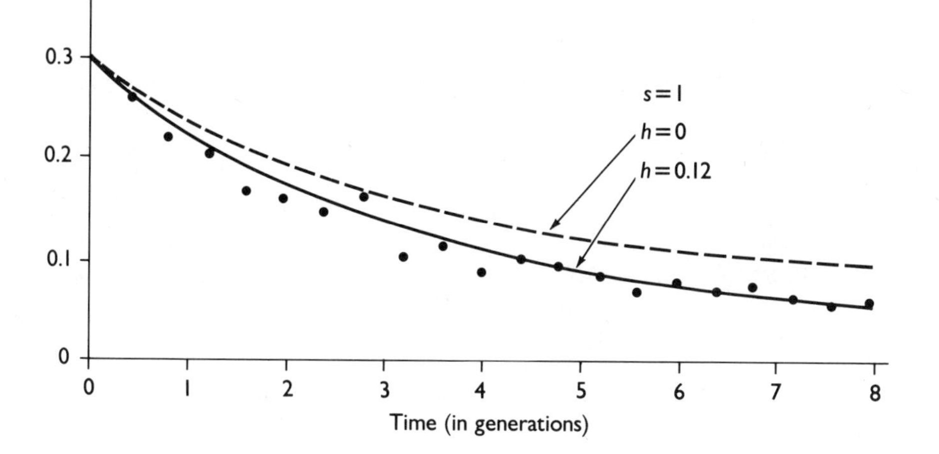

genotypes: $+/+$ (normal), $+/Sb$ (stubble bristles), and Sb/Sb, which is lethal in the larval stages and thus not seen in adults. The generation time (average age of reproduction) in these cages was about 2.5 weeks. A sample of 200 adults was classified each week. Each dot is the average of four populations started simultaneously. The dashed line is the expectation if the mutant behaves as a recessive lethal ($s = 1$, $h = 0$). Clearly this line does not fit the data. The solid line, which fits very well, is for $s = 1$, $h = 0.12$.

The real population had eggs, larvae, pupae, and adults all coexisting in a cage. The generations were overlapping. The environment was relatively constant, but certainly not absolutely so; for example, the population size fluctuated considerably. Selection was strong. Yet a simple model based on random selection from a gene pool at discrete time intervals accurately predicts what actually happened.

The calculation of the changes of gene frequency over many generations is easy but exceedingly tedious unless programmed for a computer. Table 4-4 gives some examples for $s = 0.01$. Instead of giving the amount of allele frequency change over a specified number of generations, the table shows the number of generations required for a specified allele frequency change. The reason for this approach will soon be clear.

Figure 4-4 depicts the general course of selection. This figure and Table 4-4 illustrate the same principles. Allele frequency change is very slow when the recessive allele is rare, whether the selection is for or against the allele. On the other hand, when the dominant allele is rare, the change is relatively rapid. In general, the time required to change an allele frequency from near 0 to near 1 is much less if the allele is semidominant than if it is either dominant or recessive.

It is easier to see these relations if we look at Equation 4-5b, which shows why the rate of change is so slow when the recessive allele is rare. If $h = 0$, the numerator of

TABLE 4-4

The number of generations required to change the allele frequency when the selection intensity, s, is 0.01.

GENE FREQUENCY		DOMINANT	RECESSIVE	NO DOMINANCE
FROM	TO	FAVORED ($h = 0$)	FAVORED ($h = 1$)	($h = 1/2$)
0.001	0.01	232	90,231	462
0.01	0.10	250	9240	480
0.10	0.25	132	710	220
0.25	0.50	177	310	220
0.50	0.75	310	177	220
0.75	0.90	710	132	220
0.90	0.99	9240	250	480
0.99	0.999	90,231	232	462

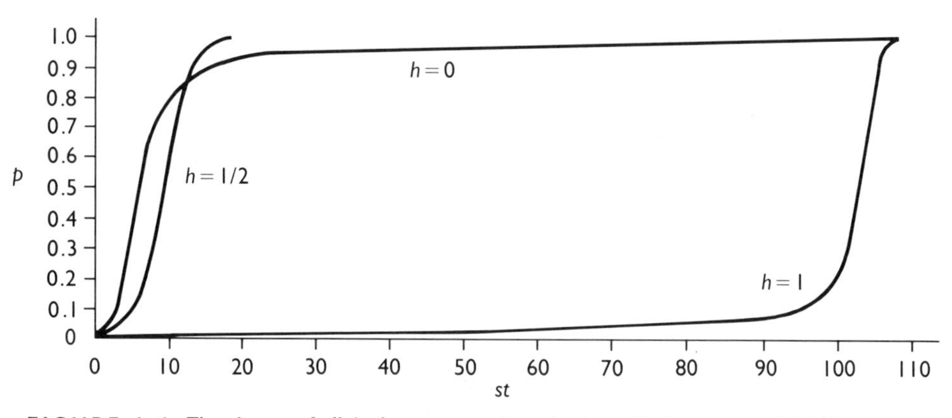

FIGURE 4-4. The change of allele frequency under selection with three types of inheritance: no dominance ($h = 1/2$), dominant favored ($h = 0$), and recessive favored ($h = 1$); $p_0 = 0.01$.

the right side of the equation is spq^2, which becomes very small when p approaches 1 and q approaches 0. In other words, the frequency of a rare recessive changes very slowly. The process is slow whether the rare recessive is favored or opposed. If the favored allele is recessive, $h = 1$ and the numerator is sp^2q, which is very small when p is small (and q large).

An Approximate Treatment for Weak Selection. If s is small, $\overline{w}$ is nearly 1. We can also replace the finite difference equation by a differential equation, as if the change were continuous. This adjustment modifies Equation 4-5b to read

$$\frac{dp}{dt} \approx spq[q + h(p - q)]. \qquad \text{4-6}$$

When written in this form, the equation is amenable to standard treatment by the calculus. I shall illustrate it for the simplest case, when there is semidominance ($h = 1/2$). In this case we can write the continuous approximation

$$\frac{dp}{dt} \approx \frac{sp(1-p)}{2}. \qquad \text{4-7}$$

Separating the variables for integration gives

$$s \int_0^t dt = 2 \int_{p_0}^{p_t} \frac{dp}{p(1-p)},$$

which after integrating gives

$$t = \frac{2}{s} \ln \frac{p_t(1 - p_0)}{p_0(1 - p_t)}. \qquad \text{4-8}$$

In Equation 4-8, t is the time required to change the allele frequency from p_0 to p_t. For example, the time required to change 0.01 to 0.99 when $s = 0.01$ is (2/0.01) ln [(0.99)(0.99)/(0.01)(0.01)], which is 1838.05. The exact answer, obtained by adding 480, 220, 220, 220, 220, 220, and 480 from Table 4-4, is 1840. Therefore the approximation works very well for weak selection.

In Equation 4-8 I could equally well have written p_t as a function of t. I chose to write t as a function of p to bring out another point. Notice that the time is inversely proportional to s. Hence if it takes 1840 generations to effect this change when $s = 0.01$, it will take twice as long when $s = 0.005$. This principle holds for any level of dominance. You can see that dominance does not affect the principle by noting that when $\bar{w}$ is 1 and we treat the process as continuous, we can write Equation 4-6 as

$$dt = \frac{(1/s)dp}{pq[q + h(p - q)]}.$$

When integrated, the constant factor $(1/s)$ remains; so t is always inversely proportional to s regardless of the value of h. We have the following simple rule:

With weak selection, the time required for a specified allele frequency change is inversely proportional to the intensity of selection as measured by s.

For example, if we would like to know how long it takes to change the frequency of a favored dominant allele from 0.10 to 0.5 when $s = 0.1$, we note first, from Table 4-4, that the time when $s = 0.01$ is 132 + 177, or 309 generations. The time when s is 10 times as large is 1/10 as long, or 31 generations.

Notice that the dominant favored and recessive favored columns of Table 4-4 are the same if one of them is read from the bottom up. This symmetry tells us, for example, that if it takes 9240 generations to change a dominant gene from 0.90 to 0.99, it also takes 9240 generations to change it from 0.99 to 0.90 if the direction of selection is reversed, since this time period is necessary to change a recessive from 0.01 to 0.10.

Most evolutionary changes, as inferred from paleontological record, are very slow. The rates are consistent with very small average values of s, 0.001 or less. Yet there are some instances of very rapid evolution, often involving human intervention, intended or unintended.

In the nineteenth century the industrial revolution spread very rapidly through parts of England. The amounts of smoke and soot rose correspondingly. Some of the white lichens that grew on tree trunks are very sensitive to smoke and were killed, thus leaving the dark-colored bark. In addition, the smoke itself helped darken the bark. During this period, some of the moths inhabiting these trees rapidly changed from a speckled light color to a much darker form. A single dominant allele causes the color difference. The frequency of this allele changed in some areas from less than 0.01 to more than 0.90 in about 35 years. This moth has only one generation per year; so this increase represents a change from about 0.01 to 0.90 in 35 generations. From Table 4-4 we can see that if $s = 0.01$, this change would require $250 + 132 + 177 + 310 + 710$, or 1579 generations. A change of this magnitude in 35 generations implies a selection intensity $s = (1579/35)(0.01) = 0.45$. Although this approximation is crude, since we are far outside the range in which the table's accuracy can be guaranteed, it gives us a rough idea and is as accurate as the data warrant. Clearly the selective intensity during this period was very high, approximately $s = 1/2$. The selective agency is mainly, if not entirely, birds. The birds were directly observed picking dark moths from light-colored trunks and light moths from dark, lichen-free trunks.

Another example of rapid selection is the increase in insecticide resistance. The first recorded instance was in 1914, when lime sulfur spray ceased to be effective against the San Jose scale. DDT and organic phosphates were widely used after World War II. By the end of 1980, 428 species of arthropods were reported to be resistant. In 1963, after 15 years of DDT spraying in Sri Lanka, there were only 17 cases of malaria. By 1968 resistant Anopheles mosquitoes were prevalent, and over a million cases of malaria were reported. There are not likely to be any clever, easy solutions. Resistance is acquired by natural selection, and such a fundamental principle cannot easily be overturned.

After the first appearance of resistant strains of insects, many observers were puzzled by the slow initial rate of increase in the number of resistant insects and the rapid increase later. This observation should not surprise you. Looking at Table 4-4 or Figure 4-4, we can see that when an allele is rare, its frequency changes very slowly. The rate is much faster as it becomes common, say 0.1. The initial rate of increase in resistance is extremely slow if the resistance allele is recessive, but is quite slow even if the allele is dominant or partially so.

As a third example of rapid gene frequency change, seasonal changes occur in some species. For example, the frequencies of some inversion types in *Drosophila pseudoobscura* change sharply from summer to winter.

Paleontological changes observed over very long time periods often show grossly different rates at different times in the same lineage. There are sometimes long periods of no observable change interspersed with seemingly instantaneous changes. Yet what looks like an instant to a paleontologist may be many generations, a long time to a population geneticist. There is currently considerable discussion as to whether evolution occurs by long periods of stasis and sudden jumps ("punctuated equilibrium") or whether the jumps are artifacts of an incomplete fossil record. We shall return to this topic in Chapter 7.

4-3 MULTIPLE ALLELES AND THE AVERAGE EXCESS OF AN ALLELE

We can write Equation 4-4 in another form that is often used in the literature. The basic idea was introduced by R. A. Fisher in his book *The Genetical Theory of Natural Selection* (1930, 1958).

Using the symbols of Table 4-2, we can rewrite Equation 4-4 as

$$\Delta p_1 = \frac{p_1(w_1 - \overline{w})}{\overline{w}}, \tag{4-9}$$

where

$$w_1 = \frac{p_1^2 w_{11} + p_1 p_2 w_{12}}{p_1} = p_1 w_{11} + p_2 w_{12}. \tag{4-10}$$

The quantity w_1 is the average fitness of all individuals carrying the A_1 allele, assigning to each heterozygote carrying this allele a weight of 1/2 because only half of its alleles are A_1. As before, $\overline{w}$ is the average fitness of the entire population. The quantity $w_1 - \overline{w}$ is called the **average excess** of the allele A_1.

The average excess is the amount by which the weighted fitness of A_1-containing individuals exceeds that of the population as a whole. Not surprisingly the rate of change of an allele is proportional to its average excess.

Remember, though, that the average excess of an allele is not a constant, even if the genotypic fitnesses are regarded as constant, since it depends on the relative frequency of homozygotes and heterozygotes. Equation 4-9 is very useful because it tells us at any generation whether an allele is increasing or decreasing. If the average excess of an allele is positive, it is increasing, and vice versa.

We can easily generalize this equation to accomodate multiple alleles. I shall not derive it explicitly, since I think it will be clear to you intuitively. The average excess of allele A_i is the average fitness of all genotypes carrying this allele, with the homozygote given a weight of 1 and each of the heterozygotes a weight of 1/2. This definition leads directly to the following equation:

$$\Delta p_i = \frac{p_i(w_i - \overline{w})}{\overline{w}}, \tag{4-11}$$

where

$$w_i = p_1 w_{i1} + p_2 w_{i2} + p_3 w_{i3} + \ldots = \sum_j p_j w_{ij}$$

$$\overline{w} = \sum_i \sum_j p_i p_j \, w_{ij}.$$

In many cases we are interested in one particular allele and not in the number of other alleles. Often the other alleles are equally, or approximately equally, fit. In this case we can simply treat all the other alleles as if they were one allele, with a frequency equal to their sum. In the following sections I shall follow this procedure, often without mentioning it explicitly. But first we shall consider the rate at which fitness itself changes under natural selection.

4-4 FISHER'S FUNDAMENTAL THEOREM OF NATURAL SELECTION

Let us now again regard the fitnesses, as measured by the w's, as absolute rather than as relative quantities. Thus w_{ij} is half the number of descendants in the next generation from an A_iA_j individual. The half comes from sexuality; an average parent in a nongrowing population has two descendants, each sharing half of each parent's genes. Now let us see how the average value of w changes.

Noting that $p_1' = p_1 + \Delta p$, we can write the mean fitness in the next generation as

$$\begin{aligned}\bar{w}' &= (p_1 + \Delta p)^2 w_{11} + 2(p_1 + \Delta p)(p_2 - \Delta p) w_{12} + (p_2 - \Delta p)^2 w_{22} \\ &= \bar{w} + 2\Delta p\,(p_1 w_{11} + p_2 w_{12} - p_1 w_{12} - p_2 w_{22}) \\ &\quad + (\Delta p)^2 (w_{11} - 2w_{12} + w_{22}).\end{aligned} \tag{4-12}$$

Recall from Equation 4-10 that $w_1 = p_1 w_{11} + p_2 w_{12}$ and $w_2 = p_1 w_{12} + p_2 w_{22}$. Substituting these expressions into the middle expression on the right side of Equation 4-12, assuming that $(\Delta p)^2$ is small enough to be neglected and noting that $\Delta\bar{w} = \bar{w}' - \bar{w}$, we have the following approximate formula:

$$\Delta\bar{w} \approx 2\Delta p(w_1 - w_2). \tag{4-13}$$

Now subtract $\bar{w}$ from w_1 and w_2, yielding

$$\Delta\bar{w} \approx 2\Delta p[(w_1 - \bar{w}) - (w_2 - \bar{w})]. \tag{4-14}$$

Now note that $\Delta p = \Delta p_1$. Note also that an increase in p_1 corresponds to the same decrease in p_2, so that $\Delta p_1 = -\Delta p_2$. Making these substitutions in Equation 4-14 yields

$$\Delta\bar{w} \approx 2\Delta p_1 (w_1 - \bar{w}) + 2\Delta p_2 (w_2 - \bar{w}). \tag{4-15}$$

Finally, if we substitute from Equation 4-9 and divide both sides of the equation by $\bar{w}$, we have

$$\frac{\Delta \bar{w}}{\bar{w}} \approx \frac{2[p_1(w_1 - \bar{w})^2 + p_2(w_2 - \bar{w})^2]}{\bar{w}^2}. \qquad 4\text{-}16$$

The generalization to three or more alleles is straightforward; we simply have three or more terms instead of two. The expression in brackets is the variance, V, of the w_i's (see Table A-1 in the appendix); that is, it is the variance of the average excesses of the different alleles at this locus. Thus we have the approximation

$$\frac{\Delta \bar{w}}{\bar{w}} \approx \frac{V(w)}{\bar{w}^2}. \qquad 4\text{-}17$$

The variance of the average excesses is called the **additive genetic variance.** Since the variance is measured in squared units, it is appropriate to standardize it by dividing by $\bar{w}^2$. Thus we can state one form of Fisher's fundamental theorem as follows:

> **The relative (geometric) rate of increase in mean fitness in any generation is approximately equal to the standardized additive genetic variance of fitness at that time.**

The rate at which selection changes any property of a population obviously depends on how much genetic variability for this property exists in the population. However, what is not obvious is that the proper units by which to measure variability are squared deviations. Fortunately the measure of variability, the variance, that statisticians have long used turns out to be the appropriate measure for computing the effect of natural (or artificial) selection on a population.

Note that the additive genetic variance is not the variance of the fitnesses of the several genotypes; instead it is the variance of the average fitnesses of the alleles. This system of fitness cost accounting allows for Mendelian segregation, since the average value of the alleles, not of the genotypes, is what is transmitted to the next generation.

As discussed in the next chapter, we can predict the rate of change of any character provided we know the correlation of this trait with fitness. If large animals are more viable or fertile, size will increase at a rate predictable by the additive genetic variance of size and the genetic correlation of size with fitness. We can use this theorem to derive the formulae that animal and plant breeders use to predict the expected improvement from artificial selection experiments (see C&K, pages 255 ff).

This single-locus, random-mating, discrete-generation special case does not do justice to the generality of the Fisher theorem. Fisher pointed the way to an understanding of how the rate of change in fitness (or other traits) can be understood even when the various complications of multifactorial inheritance — dominance, epistasis, linkage — are present. He showed that natural selection, although it acts on phenotypic differences, picks out the additive genetic component and changes the population mean in proportion to the variance of this component.

One aspect of this theorem is puzzling at first glance. If we take the theorem literally, the mean fitness of the population should increase indefinitely as long as there is additive variance for this trait. Clearly the fitness cannot increase indefinitely; the world is finite. In fact, the mean fitness, $\overline{w}$, of most species over any long period of time must be very close to 1.

One way to resolve this issue is by the following argument. The genetic capacity for survival and fertility does indeed increase in each species. Yet the environment is deteriorating, which offsets the genetic improvement. For any one species the environment tends to get worse because of diminishing resources, increasing waste products, and other problems of overcrowding. Perhaps even more important, all competing species are improving their fitnesses too and thus making life more difficult for one another. Each species is like Alice on the treadmill: it has to keep running to stay in the same place. Hence, $\overline{w}$ remains nearly constant.

Evolution in a closed environment is vividly illustrated by chemostat experiments in bacteria. A chemostat is a steady-state growth chamber in which a liquid culture medium is pumped in at a constant rate. An overflow tube allows both media and bacteria to flow out. The bacteria come to equilibrium with a bacterial multiplication rate that exactly balances the outflow. Mutants occur, and those making their bearers better adapted to chemostat life will successively replace their predessors. Yet the total density and growth rate of the bacterial colony hardly change, because they are governed by the concentration and flow rate of the medium. Nevertheless a bacterial strain that has been in a chemostat for several days and has accumulated several mutants would quickly replace the ancestral strain from which it arose if the two were placed in direct competition in a chemostat. For a discussion of the use of a chemostat for evolution experiments, see Hartl (1980).

You might enjoy reading Fisher's own formulation of the fundamental theorem in his book, *The Genetical Theory of Natural Selection* (1930,1958). You will probably find it difficult; most people do. A more straightforward, though less elegant, derivation is given in C&K, pages 205 ff.

4-5
SELECTION BALANCED BY MUTATION

From the previous section you learned that consistent, directional selection changes the allele frequencies, sometimes quite rapidly but more often slowly. Eventually those alleles that are most favorable — those with the highest average excess — will replace the less favorable alleles, although this process may take a very long time. Yet whenever we observe a natural population, many deleterious alleles are always present. The usual explanation is that they arose by mutation and have not yet been eliminated by natural selection.

We shall consider the equilibrium allele frequencies when elimination of deleterious alleles is balanced by new mutations.

Equilibrium Between Mutation and Selection. The general model is the same as that shown in Table 4-3, except that we want to include mutation. Equation 4-5a gives the frequency of allele A, p', in terms of its frequency, p, in the previous generation. The allele will have remained unchanged only if it has not mutated. If μ is the mutation rate from A to A' per gene per generation, then the probability that allele A has remained unchanged is $1 - \mu$. We modify Equation 4-5a accordingly:

$$p' = \frac{p(1 - qhs)}{\overline{w}}(1 - \mu), \qquad \text{4-18}$$

where

$$\overline{w} = 1 - sq(q + 2ph).$$

To obtain the equilibrium value, we follow the following familiar practice: setting $p' = p$ and solving the resulting equation for $q = 1 - p$. This leads to the quadratic equation

$$s(1 - 2h)\hat{q}^2 + hs(1 + \mu)\hat{q} - \mu = 0, \qquad \text{4-19}$$

in which I have again used a circumflex to designate the equilibrium value.

I have ignored reverse mutation from A' to A. This omission does not lead to any significant error for two reasons: first, reverse mutation is almost always less than the forward rate by an order of magnitude or more. Second, if s is much larger than μ, which is true for the cases we are considering, the mutant allele is rare and therefore only a few mutant genes are available to mutate back to normal.

We can solve Equation 4-19 by the standard quadratic formula, but first let us consider three special cases that encompass most of the values of interest.

Recessive mutant,* h = *0. In this case Equation 4-19 reduces to

$$\hat{q}^2 = \frac{\mu}{s} \qquad \hat{q} = \sqrt{\frac{\mu}{s}}. \qquad \text{4-20}$$

For example, if the mutation rate from A to A' (which I shall designate as a to indicate its recessiveness) is $1/100{,}000$ and $s = 1/10$, the equilibrium value of q^2 is

1/10,000 and $q = 1/100$. The general rule is easy to remember because it is so reasonable.

The proportion of homozygous recessive genotypes at equilibrium is the ratio of the mutation rate to the selective disadvantage of the mutant homozygote.

A practical difficulty in applying this rule is that the equilibrium between mutation and selection against a recessive allele is approached very slowly. Selection occurs only when the mutant is homozygous, and dozens or, in a large population, hundreds of generations may pass between such occurrences. It is unlikely that s will have remained constant for this long a period, unless $s = 1$. Another reason why Equation 4-20 is likely to be inaccurate is that the frequency of the mutant allele may fluctuate by random drift while it is heterozygous. Finally any inbreeding in the population can greatly change the equilibrium value by making recessive alleles homozygous and exposing them to selection. I emphasize all these uncertainties, lest you be tempted to use Equation 4-20 to estimate the mutation rate of a recessive allele.

Actualy the difficulty is not as great as it may seem, for very few mutants are completely recessive, which leads us to partial dominance.

Semidominance,* h = *1/2. In this case Equation 4-19 reduces to

$$\hat{q} = \frac{2\mu}{(1+\mu)s} \approx \frac{2\mu}{s}. \qquad \text{4-21}$$

Again the result is intuitively reasonable: the equilibrium frequency of the allele is (approximately) the mutation rate divided by the selective disadvantage of the mutant heterozygote. You might find it interesting that this equation is exact, rather than only approximate, when the heterozygote fitness is the geometric, rather than the arithmetic, mean of the two homozygotes (see Problem 13 at the end of this chapter).

Partial Dominance: A Useful Approximation. If $h^2 s \gg \mu$, the mutant allele will be rare in the population, and therefore q^2 is very small. If we neglect the term containing q^2 in Equation 4-19, we easily obtain the approximation

$$\hat{q} \approx \frac{\mu}{hs}. \qquad \text{4-22}$$

We would expect this equation to be accurate except near the border line of complete recessivity. As $h^2 s$ approaches 0, the equilibrium value of the mutant allele changes from μ/hs to the square root of μ/s. During the transition, neither equation is

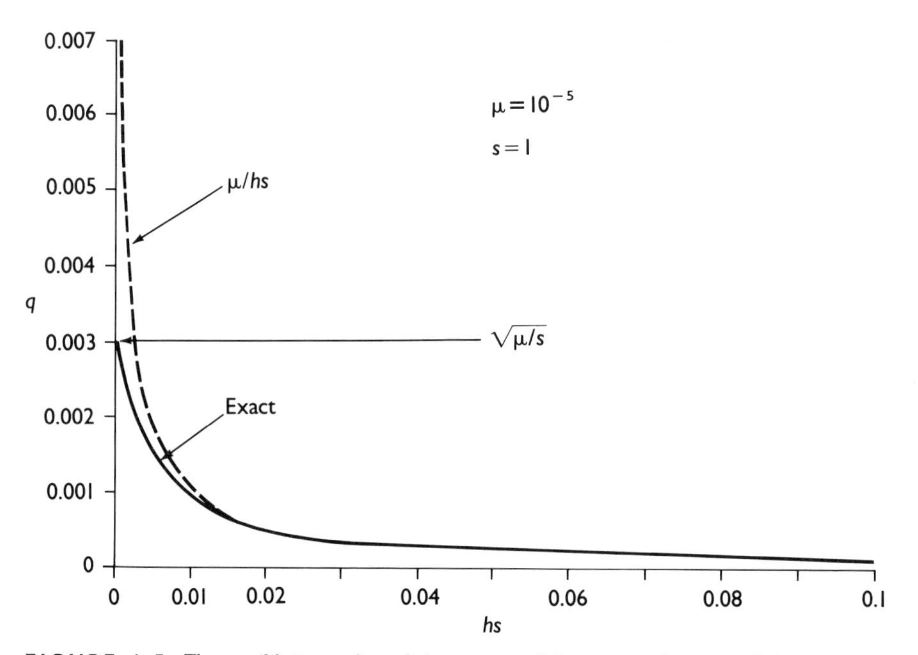

FIGURE 4-5. The equilibrium value of the mutant allele, q, as a function of the heterozygous disadvantage, hs. The mutation rate, μ, is 10^{-5}, and $s = 1$. The dashed line shows the approximation $\hat{q} = \mu/hs$; the solid line is the exact value, computed by solving the quadratic Equation 4-19.

accurate. Figure 4-5 shows this situation graphically. In this example the approximation is quite accurate until hs becomes 0.01 or less. The more precise rule (which I shall not derive here) is that the formula is accurate as long as h^2s is an order of magnitude larger than μ. In this example $\mu = 10^{-5}$ and $s = 1$; so we would expect the formula to be exact whenever $h^2s \gg 0.00001$, or $hs \gg 0.003$, say $hs > 0.03$ as the graph verifies. For borderline cases we can always solve the quadratic Equation 4-19.

Two Examples of the Practical Use of Equation 4-22. A standard method in human genetics is to use Equation 4-22 to compute mutation rates. A good example is achondroplasia, or short-limbed dwarfism. A classical study found an incidence of 10 achondroplastics among 94,075 births. They were all heterozygotes; so the mutant allele frequency, q, is half the incidence, or 5/94,075. To estimate s, we note that 108 achondroplastics left 27 descendants to the next generation, whereas 457 sibs produced 582 descendants. The relative fitness of achondroplastics, $1 - hs$, is $(27/108)/(582/457) = 0.20$. Thus $hs = 0.8$, and using Equation 4-22, we have $\mu = (0.80)(5/94{,}075) = 4.3 \times 10^{-5}$.

This value agrees well with the direct estimate, obtained by counting the number of affected children from normal parents. In this case eight of the affected newborns were new mutants. Thus the mutation rate is estimated as $(8/94{,}075)/2 = 4.3 \times 10^{-5}$. We divide by 2 because the new mutation could have been in either the egg or the sperm. The exact agreement in the two estimates, which are both subject to large errors, is fortuitous.

The indirect method using Equation 4-22 has the advantage that the parentage need not be known, and therefore it avoids the frequent complication of errors in determining paternity. Using both methods provides a useful check.

As a second example we can use Equation 4-22 to learn something about the dynamics of lethal genes in Drosophila. The mutation rate is 0.005 per chromosome generation for each of the major autosomes. Since each chromosome contains about 2000 lethal-producing loci, the mutation rate per locus, μ, is $0.005/2000 = 2.5 \times 10^{-6}$. When chromosomes are extracted from natural populations and made homozygous by the standard method of using genetically marked chromosomes with recombination-suppressing multiple inversions, the mean number of lethals per chromosome is about 0.25. Thus the lethal allele frequency, q, is $0.25/2000 = 1.25 \times 10^{-4}$.

If the lethals were completely recessive, we would expect $\hat{q}$ to be the square root of the mutation rate (from Equation 4-20 with $s = 1$), which is 1.58×10^{-3}. This value is 13 times the observed value, indicating that the assumption of complete recessivity of the lethals is wrong. Instead let us assume partial dominance and use Equation 4-22. Since $s = 1$, $hs = h = \mu/q = 0.02$. Thus the data are consistent with the "recessive" lethals being partially dominant, with the heterozygous effect of the mutant being about 2 percent of the homozygous effect.

Although we cannot be as specific about lethal mutation rates in the human population, qualitatively they seem very similar to those of Drosophila. If the lethal mutation rate is 10^{-6}, the equilibrium frequency of a completely recessive allele is 10^{-3}. No one knows how many lethal-producing human loci there are, but the number is probably at least as large as that in Drosophila. Let us say, as a guess, that there are 10,000 per haploid genome (about twice the Drosophila number). The number of lethal equivalents per gamete would then be about 10. If the mutation rate is higher or the number of loci is larger, this number would be greater than 10. On the other hand, estimates of the number of lethal equivalents per gamete, obtained from death rates in children of consanguineous marriages, are closer to 1 than to 10, although we are undoubtedly missing some that act prenatally. Therefore as with Drosophila, we have a discrepancy, with the most likely explanation being partial dominance.

Studies of spontaneous mutants in Drosophila have shown that h and s are negatively correlated. In fact, if we take the data at face value, hs seems to be nearly constant. Since the impact of mutations on population fitness is almost entirely through heterozygotes, a constant hs means that mutants whose homozygous effects are mildly deleterious cause as much harm to the population as do those whose homozygous effects are drastic. Since mild mutants occur much more frequently, they make a much larger impact on population fitness.

What is the implication of this finding for human welfare? How do we compare a large number of people that are slightly weakened by mildly deleterious heterozygous effects with a much smaller number of painfully diseased homozygotes? Difficult as it is to say anything quantitatively reliable, the cumulative effect of minor mutations is substantial and is one reason to try to keep the total mutation rate as low as possible.

The Haldane Mutation Load Principle. In 1937 the British polymath J. B. S. Haldane asked the question, How much does mutation impair the average fitness of the population? For a single locus we can easily compute this value.

Note that for a recessive mutation the frequency of mutant homozygotes is μ/s. The effect of the mutant is to lower homozygous fitness by a factor s. Hence the reduction in fitness due to mutation at this locus is $(\mu/s)s$, or simply μ.

Likewise, with partial dominance the number of mutant heterozygotes is approximately $2q = 2\mu/hs$. Each mutant heterozygote has its fitness reduced by a factor hs, so that the net effect is $(2\mu/hs)hs$, or 2μ. From this type of analysis Haldane said that if we add up the effects of all loci, the impact of mutation on population fitness is equal to the gametic mutation rate multiplied by a factor of 1 or 2 depending on dominance. Since partial dominance is clearly the rule, the effect of mutation on the population, called the **mutation load** (after H. J. Muller, who independently discovered the Haldane principle), is twice the total mutation rate per gamete.

This principle undoubtedly works well for mutants with large heterozygous effects. It tells, for example, that if the total mutation rate per gamete for such loci is 0.01, the population fitness is lowered by twice this amount. If the mutation rate were doubled, the effect would double after the new equilibrium is reached, and the population fitness would be lowered by a factor of 0.04.

For mild mutations the principle is of questionable accuracy. The relationship between the mutation rate and mutation load is distorted because the effects of multiple mutations are usually worse than if they were independent. One premature death or failure to reproduce often removes several alleles at once and reduces the effect of each. Thus where mild mutants are involved, the mutation load is less, possibly much less, than the Haldane principle would predict.

One way to incorporate mild mutations into the formula is as follows: the mutation load is equal to the mutation rate per zygote (twice the rate per gamete) divided by the difference between the average number of mutant genes in the individuals eliminated by selection and in those left to reproduce. This formula shows that if nature or the breeder can contrive to eliminate those individuals with the greatest number of mutations, the mutation load will be reduced correspondingly. Truncation selection, in which individuals below a fitness threshold are eliminated while all above the threshold are retained, maximizes the efficiency of mutant elimination. As you will see in Chapter 5, this is what an animal or plant breeder does. Whether or not nature truncates is an open question, but it may often come close to doing so.

Regardless of the complexities of calculating the mutation load, the impact of mutation on the fitness of the population is proportional to the mutation rate and not to the selective disadvantage of the individual mutant alleles.

4-6 SELECTION AT AN X-LINKED LOCUS

Analysis of X-linked loci is complicated by the fact that we cannot assume that the two sexes have the same allele frequencies and therefore we cannot realistically assume Hardy-Weinberg ratios. We will concentrate on mutations with a large, deleterious effect, so that the fitness of affected males is greatly reduced. Therefore we can ignore the homozygous mutant females, which will be very rare. The model is set forth in Table 4-5.

Following our custom of writing equations for successive generations and using a prime to designate the next generation, we can write the equation for affected males in terms of the proportion of carrier females in the preceding generation. If the frequency of heterozygous females is H, then half the time the allele inherited by a son will be mutant. In the other half of the cases, the normal allele will be inherited, but it will have mutated in the germ cells of the mother with probability μ_f. The total contribution of mutant alleles from heterozygous mothers is $H(1/2 + \mu_f/2)$. The probability that the mother is homozygous normal is $1 - H$, and the probability that the transmitted gene mutated is μ_f. Putting all this information together, we have

$$\begin{aligned} q' &= \frac{H}{2} + \frac{H\mu_f}{2} + (1 - H)\mu_f \\ &\approx \frac{H}{2} + \mu_f. \end{aligned} \tag{4-23}$$

The approximation comes from neglecting any term involving the product of the two small fractions, μ_f and H.

TABLE 4-5

A model of a strongly selected X-linked locus. Mutation rate from A to a: μ_m in males; μ_f in females.

	FEMALES			MALES	
GENOTYPE	*AA*	*Aa*	*aa*	*A*	*a*
Relative fitness	1	1	<1	1	$1 - s$
Frequency at birth	$1 - H$	H	0	p	q

In the same way we obtain a corresponding equation for carrier females. This time we have to consider that the mutation could also happen in the male parent. Adding the female and male contributions gives

$$\begin{aligned} H' &= \frac{H}{2} + \frac{H\mu_f}{2} + (1 - H)\mu_f + q(1 - s) + (1 - q)\mu_m \\ &\approx \frac{H}{2} + \mu_f + q(1 - s) + \mu_m. \end{aligned} \tag{4-24}$$

At equilibrium $H' = H$ and $q' = q$. Substituting $q - \mu_f$ from Equation 4-23 for $H/2$ in Equation 4-24 yields the equilibrium equation

$$\hat{q} \approx \frac{\mu_m + 2\mu_f}{s}. \tag{4-25}$$

Substituting this expression into Equation 4-24 gives

$$\hat{H} \approx \frac{(2\mu_m + 4\mu_f)(1 - s)}{s} + 2\mu_f + 2\mu_m. \tag{4-26}$$

If the mutation rates are the same in the two sexes and if we have an estimate of fitness of affected males, we can use Equation 4-25 to estimate the mutation rate.

X-linked mutants offer a unique opportunity to compare human male and female mutation rates. For diseases such as hemophilia, Lesch-Nyhan syndrome, and Duchenne muscular dystrophy, s is essentially 1. Note from Equation 4-23 that the frequency of males affected from new mutants is μ_f, whereas the frequency from carrier mothers is $H/2$, which at equilibrium is $\mu_m + \mu_f$. Thus if the two rates are equal, 1/3 of the affected males are new mutants. Otherwise, among affected males, the

$$\text{Proportion of new mutants} = \frac{\mu_f}{\mu_m + 2\mu_f}. \tag{4-27}$$

For hemophilia only about 1/12 of affected males are new mutants. If we let the male rate, μ_m, equal $k\mu_f$, substitute the expression into Equation 4-27, and solve for k, we get $k = 10$. This value of k implies that mutation rates are an order of magnitude higher in males than in females, which is reasonable in view of the much larger number of cell divisions in the male germ line.

4-7 OPPOSING SELECTION FORCES

Selection on an allele may work in different directions in different genotypes, such that the same gene is favored under some circumstances and not under others. For example, a gene that increases size will be advantageous in small individuals but deleterious in large ones, since the optimum size is intermediate. At a specific locus genotype AA' may be more fit than either homozygote, which is sometimes called **overdominance.** Thus an A allele is advantageous when coupled with an A' partner but not with an A partner.

Comparable situations can arise when two or more loci are involved. For example, maternal-fetal incompatibility for the ABO blood groups decreases the risk of Rh hemolytic disease. If we all had the same ABO blood group, Rh disease would be more prevalent. Also, the presence of a polymorphic blood group locus in apes suggests other advantages to retaining all three alleles. Haldane and others have proposed that the blood group polymorphism is the result of some disease-resistance mechanism. Possibly it is a relic of a disease, such as smallpox or plague, that is no longer the major killer it once was.

Equilibrium with Overdominance for Fitness. This situation is best discussed with an example. The abnormal hemoglobin, S, makes the red blood cells inhospitable to the malaria parasite *(Plasmodium falciparum)*. On the other hand, persons with only this kind of hemoglobin have sickle-cell anemia. The model is set forth in Table 4-6.

I have used the heterozygote as a reference point, with relative fitness 1, since it has the highest rates of survival and fertility. Again we express next generation's allele frequency in terms of the current frequency, which leads to

$$p' = \frac{p^2(1-t) + pq}{p^2(1-t) + 2pq + q^2(1-s)} = \frac{p(1-tp)}{1 - tp^2 - sq^2}. \qquad \text{4-28}$$

TABLE 4-6

Selection at a locus where the heterozygote has greater fitness than either homozygote. The example is taken from hemoglobin types.

GENOTYPE	*AA*	*AA'*	*A'A'*
Hemoglobin type	A	A and S	S
Phenotype	"Normal"	Malaria resistant	Anemic
Fitness	$1-t$	1	$1-s$
Frequency at birth	p^2	$2pq$	q^2

Letting $p' = p$ and solving the equations, we have

$$\hat{p} = \frac{s}{s + t} \qquad \hat{q} = \frac{t}{s + t}. \tag{4-29}$$

These equations show that when $p = s/(s + t)$, the frequency of the allele has no tendency either to increase or decrease. But is the equilibrium stable? In the case of mutation-selection balance, I took it for granted that the equilibrium was stable in the sense that if for any reason the allele frequency departed from the equilibrium value, selection would tend to carry it back toward the equilibrium. The same condition is true in this case. If we start with a value on one side of $s/(s + t)$ and apply Equation 4-28, the frequency in the next generation will be closer to the equilibrium value. Hence the allele frequency tends to move toward the equilibrium value, and the equilibrium is stable.

We can easily show the movement toward equilibrium algebraically by using the concept of average excess. The average excess of allele A, $w_A - \overline{w}$ (using the definitions of Equation 4-9), is

$$\begin{aligned} E_A &= p(1 - t) + q - (1 - tp^2 - sq^2) \\ &= -q\,[p(s + t) - s]. \end{aligned} \tag{4-30}$$

Notice that if p is greater than the equilibrium value, $s/(s + t)$, the excess is negative; therefore the allele frequency decreases. If p is below the equilibrium value, the excess is positive and the frequency increases. Thus the allele frequency tends to move toward the equilibrium value.

You can now understand why the gene for sickle-cell anemia is frequent in those parts of the world where malaria has been a major killer. The allele increases up to its equilibrium value, despite the anemia in sickle-cell homozygotes. It is certainly not as efficient as a malaria-resistant mutant that is not deleterious in homozygotes, but presumably no such mutant was available.

We can use equilibrium theory to get some idea of the impact of malaria on the population during the period when this equilibrium was reached. Let us assume that $s = 1$, since sickle-cell anemia is usually lethal when medical and living conditions are primitive. In some parts of Africa where malaria is endemic, the frequency of the sickle-cell gene, q, is about 0.1. If $q = 0.1$ and $s = 1$, we can use Equations 4-29 to solve for t, which turns out to be 1/9. Thus the AA homozygotes are 1/9 less fit than the heterozygotes, showing that malaria has indeed been a major factor in reducing the fitness of the population. The point of interest here is that we are able to infer the average relative fitnesses of genotypes in the past despite the absence of any demographic data for that time.

Compared to heterozygotes, the total reduction in population fitness caused by *AA* homozygotes is $tp^2 = 9/100$. The reduction from sickle-cell homozygotes is $sq^2 = 1/100$. We have the paradoxical conclusion that the mildly affected homozygote causes much more reduction in population fitness than the more severely affected one.

In the past and in many parts of the world even today, malaria is the most important of all diseases. As such it has played a large role in human evolution. It has also played a role in the establishment of polymorphisms other than hemoglobin S. The global distribution of the X-linked recessive allele for glucose-6-phosphate dehydrogenase deficiency, which causes red-cell hemolysis after ingestion of certain drugs or foods (especially fava beans), coincides with malaria. So does the distribution of thalassemia, another type of anemia. In each case the price for malaria resistance is homozygotes or hemizygotes that are susceptible to anemia. Finally, the Duffy blood group allele, *Df-a*, is also associated with malaria resistance, but this time to the milder tertian malaria, caused by *Plasmodium vivax*.

You might think that a possible solution to the hemoglobin S dilemma would be for a duplication to occur, possibly by a mistake in crossing over, so that alleles A and A' occur on the same chromosome. Homozygotes for such a chromosome should produce both kinds of hemoglobin, as do present heterozygotes. A duplication of this sort would eliminate the problem of both anemic and malaria-susceptible homozygotes. Whether or not such a duplication has ever occurred or whether or not it is in some way harmful is unknown; it might possibly upset the balance between amounts of alpha and beta hemoglobins. Yet at face value it certainly appears to be a better solution to the malaria problem than the one evolution has produced.

Other instances of genetically maintained polymorphism do occur. One example was given in Chapter 1, in which Drosophila heterozygous for an inversion were shown to be more viable than either homozygote. The reason is that each inversion, by suppressing recombination, locks in some deleterious recessive alleles. Since those recessives in the inverted and standard chromosome orders are different, they are "covered" in the heterozygote. There need be no overdominance at individual loci.

Are overdominant gene loci a major mechanism by which polymorphisms are maintained? This would seem to be plausible, but it has been demonstrated in very few cases. This mechanism was once thought to be responsible for the heterozygosity of isozyme-producing loci. However, most recent evidence now points the other way. As mentioned earlier, haploid *E. coli* are as polymorphic as diploid species. The heterozygosity may well be due to very mildly deleterious mutations maintained by a mutation-selection balance.

Selection Against Heterozygotes. We can deal with selection against the heterozygote by using the model of Table 4-6 but changing the s and t to negative quantities. Then the heterozygote is less fit than either homozygote. We can carry through the algebra of Equations 4-28 and 4-29 exactly as before. The equilibrium values of p and

q, as given by Equation 4-29, are unchanged, because the minus signs in the numerator and denominator cancel. The equilibrium formula is the same whether the heterozygote is favored or the homozygotes are favored.

One thing is different, however. The equilibrium is no longer stable. If we go through the same algebra that led to Equation 4-30, we note that E_A is the same except that the minus sign to the right of the equality sign has disappeared. So if p is greater than $s/(s + t)$, it tends to become still greater, and if less, still less. If the allele frequency is displaced from the equilibrium point, the population tends to move farther away, until one or the other allele is fixed (except for mutation, which we are ignoring). Which allele is finally fixed depends on which side of the equilibrium point the initial allele frequency lay.

An unstable equilibrium is like a stone balanced on a mountain peak. It has no tendency to move either way, but if we push it slightly, it will roll down. A stable equilibrium, which occurs when the heterozygote is favored, is like a stone at the bottom of a crater. If it is displaced, it rolls back to the equilibrium rather than away from it. In the real world there is always something, random drift if nothing else, that carries the gene frequency away from an unstable equilibrium. Therefore we should not expect to find populations that are polymorphic for loci where the heterozygote is at a disadvantage. If such a situation is found, it must be transitory.

The best understood examples of selection against heterozygotes are not single loci, but chromosomal. A translocation heterozygote is at a disadvantage compared with either homozygous gene arrangement, because the heterozygote produces chromosomally unbalanced gametes. Therefore we would not expect to, and seldom do, find populations that are polymorphic for translocations. When found, such populations usually have an explanation, often a tricky one, like the set of balanced lethals in Oenothera. The vast majority of new translocations are quickly eliminated from the population. Inversions usually suffer the same fate because of unbalanced chromosomes produced by crossing over within the inverted region. Drosophila is an exception. Because of the absence of crossing over in males and a tricky way of shunting crossover chromosomes into the polar body in females, inversions are rendered nearly harmless and are often favored because, as mentioned earlier, they lock groups of genes together.

A major question in evolution is how translocations are ever established. The most likely explanation is a very small population bottleneck. In a very small population the translocation frequency may drift to a value higher than the equilibrium. In this case selection would carry the translocation on to fixation. Hence a translocation that has been fixed in evolution suggests that the population was very small at some time in the past.

A second example of selection against the heterozygote is maternal-fetal incompatibility at the Rh locus. When a fetus develops homolytic disease, it is necessarily heterozygous. For if we designate the antigen-producing gene by R, the mother must be rr or she could not produce the antibodies. The fetus must have an R gene to be antigenic and must be heterozygous since it inherited an r gene from its mother.

Under this mode of selection, except for a small number of mutations any population should be homozygous for either the R or r allele, depending on which was initially the most frequent. Most of the world's populations are indeed nearly homozygous. The Rh^- (r) allele is virtually absent in Asia and most of the rest of the world, but not in populations derived from Western Europe, where the frequency is about 0.4. How do we explain the present of both R and r alleles when theory tells us that the rarer allele should disappear? The most likely explanation is that the European population is recently of mixed origin. We know that migrants bringing R alleles have come into Western Europe from Asia and Eastern Europe. If the original European inhabitants had been Rh^- (rr), that would explain the mixed population.

There is evidence for a high Rh^- frequency in the original European population. The Basques, in southern France and northern Spain, have the highest Rh^- frequency of any part of the world, the frequency of r being greater than 1/2. For linguistic, geographical, and other reasons, anthropologists believe that these people come closest of any contemporary group to being remnants of the original European stock. The best guess is that Europeans and the populations derived from them are really relatively recent hybrids and that the selection has not had enough time to fix one allele or the other. The time scale of a few thousand years is probably insufficient for the selective elimination of one or the other allele, for the selection is weak. Hemolytic disease is rare even in a population with a high incidence of Rh^- mothers with Rh^+ children.

4-8 THREE ALLELES

We saw that in the two-allele case a polymorphism will be stable if the heterozygote is more fit than either homozygote. We also saw that no stable equilibrium could exist otherwise. Either one allele is favored and eventually fixed (except for mutation) or, if the heterozygote is less fit than either homozygote, one or the other allele is fixed, depending on which side of the equilibrium point the population started from. We are not surprised that the situation is more complicated when more than two alleles are involved.

As long as the fitnesses of the individual genotypes are constant and we assume random mating, the mean fitness of the population will always increase unless it is at equilibrium. You can find proofs of this reasonable theorem in C&K, page 273, and more succinctly and rigorously in Nagylaki (1977), pages 60 ff. In this book I shall treat only the three-allele case. The model is set forth in Table 4-7.

TABLE 4-7

A model for selection at a locus with three alleles.

GENOTYPE	A_1A_1	A_2A_2	A_3A_3	A_1A_2	A_1A_3	A_2A_3
Fitness	w_{11}	w_{22}	w_{33}	w_{12}	w_{13}	w_{23}
Frequency	p_1^2	p_2^2	p_3^2	$2p_1p_2$	$2p_1p_3$	$2p_2p_3$

The following recurrence equations are a natural extension of those we used for two alleles in, for example, Equation 4-3:

$$\begin{aligned}\bar{w}p_1' &= p_1^2w_{11} + p_1p_2w_{12} + p_1p_3w_{13}\\ \bar{w}p_2' &= p_1p_2w_{12} + p_2^2w_{22} + p_2p_3w_{23}\\ \bar{w}p_3' &= p_1p_3w_{13} + p_2p_3w_{23} + p_3^2w_{33},\end{aligned} \qquad 4\text{-}31$$

where

$$\bar{w} = p_1^2w_{11} + p_2^2w_{22} + p_3^2w_{33} + 2p_1p_2w_{12} + 2p_1p_3w_{13} + 2p_2p_3w_{23}. \qquad 4\text{-}32$$

Now set $p_1' = p_1 = \hat{p}_1$, and do likewise for p_2 and p_3. We can solve these equations by standard methods (for example, see C&K, page 275). The solutions are

$$\begin{aligned}\hat{p}_1 &= \frac{D_1}{D} \qquad D_1 = (w_{12} - w_{22})(w_{31} - w_{33}) - (w_{12} - w_{23})(w_{31} - w_{23})\\ \hat{p}_2 &= \frac{D_2}{D} \qquad D_2 = (w_{23} - w_{33})(w_{12} - w_{11}) - (w_{23} - w_{31})(w_{12} - w_{31})\\ \hat{p}_3 &= \frac{D_3}{D} \qquad D_2 = (w_{31} - w_{11})(w_{23} - w_{22}) - (w_{31} - w_{12})(w_{23} - w_{12}),\end{aligned} \qquad 4\text{-}33$$

where

$$D = D_1 + D_2 + D_3.$$

Sometimes these equations give negative values for the allele frequencies. This means that there is no equilibrium in which all three alleles are present; at least one allele will be lost. The condition for stability of an equilibrium is that the equilibrium point be a maximum for fitness. Applying the standard rules of differential calculus for a maximum and defining one more quantity,

$$T = w_{11} - w_{13} - w_{31} + w_{33}, \qquad 4\text{-}34$$

leads to the conditions

$$D_1, D_2, D_3 > 0 \qquad T < 0. \tag{4-35}$$

Figures 4-6, 4-7, and 4-8 show examples of the applications of these rules. They also show plots, drawn by a desktop computer, of the paths followed by the allele frequencies from various starting points.

In Figure 4-6 the application of Equation 4-35 shows that the D's are positive and T negative; hence the equilibrium is stable. As you can see, whatever the starting

FIGURE 4-6. The trajectories of allele-frequency change under selection, starting from various arbitrary frequencies, which are indicated by circles. The frequency of A_1 is proportional to the vertical distance from the bottom border, the frequency of A_2 the perpendicular distance from the left border, and the frequency of A_3 the perpendicular distance from the right border. The fitnesses are $w_{11} = w_{22} = w_{33} = 0.9$ and $w_{12} = w_{13} = w_{23} = 1.0$. There is an equilibrium at which each allele frequency is equal to 1/3. $D_1 = D_2 = D_3 = 0.01$, and $T = -0.2$. Thus the equilibrium is stable.

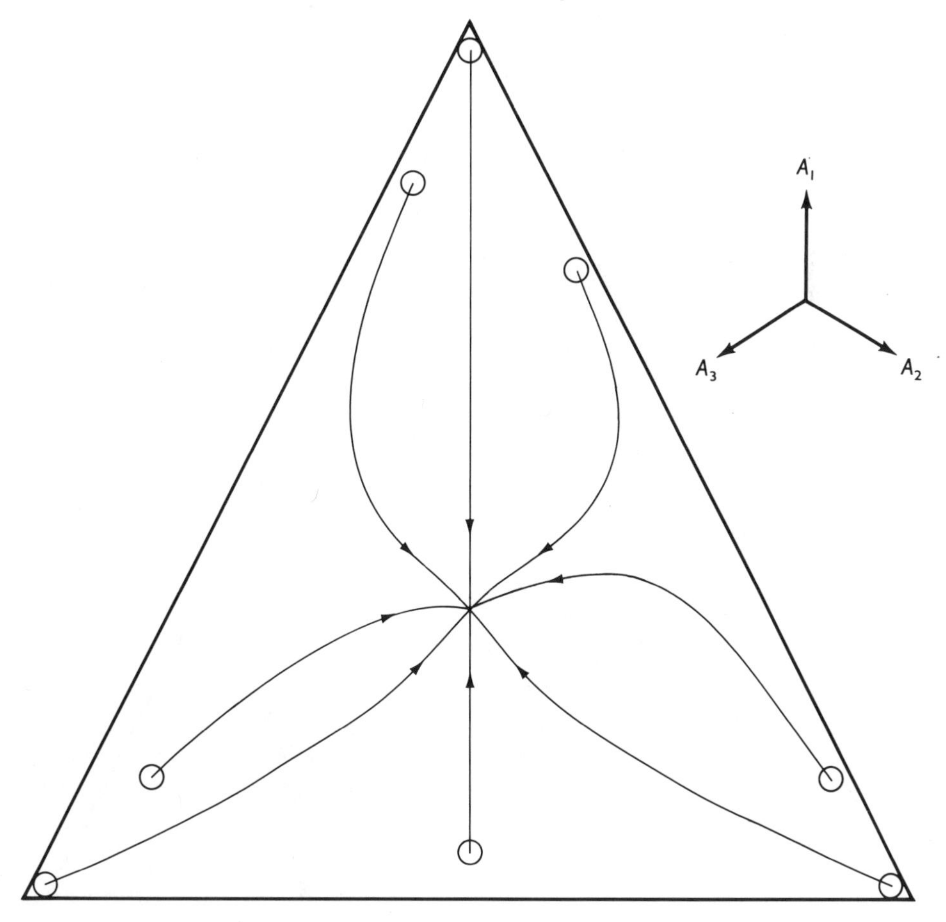

frequencies are, the population moves to an equilibrium in which all alleles are equally frequent.

In Figure 4-7 equilibrium is also reached when the allele frequencies are each equal to 1/3, but this time the equilibrium is unstable. The population moves toward homozygosity for one of the alleles, depending on the starting value.

We might have anticipated both of these results by extending our knowledge of two-allele situations. Figure 4-8 brings in something new. You might have suspected that a stable equilibrium would occur only if every heterozygote were more fit than any homozygote, but this is not required. Figure 4-8 gives an example in which the A_1A_1 homozygote is more fit than the A_2A_3 heterozygote. Yet the equilibrium is stable. Furthermore, for there to be an equilibrium, it is not sufficient for all heterozygotes to be superior to any homozygote.

FIGURE 4-7. Another set of three-allele frequency changes. The fitnesses are $w_{11} = w_{22} = w_{33} = 1$ and $w_{12} = w_{13} = w_{23} = 0.9$. Again there is an equilibrium at which allele frequency equals 1/3. However, since $T = 0.2$, the equilibrium is unstable.

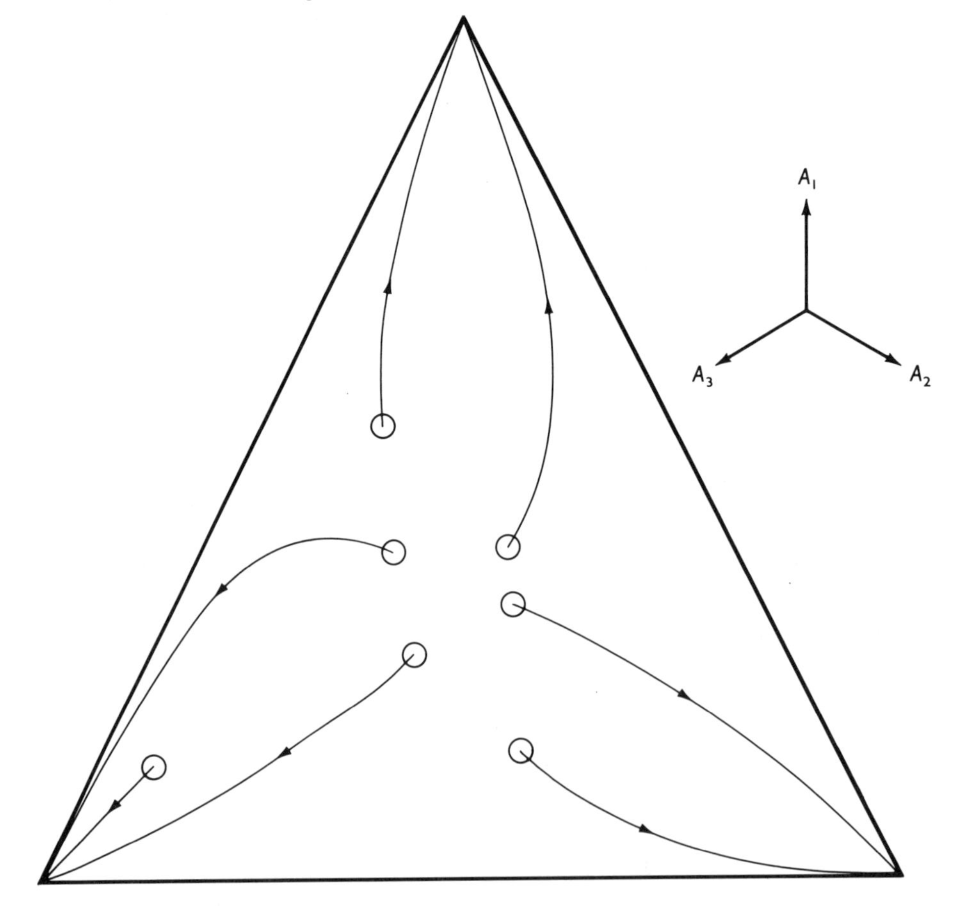

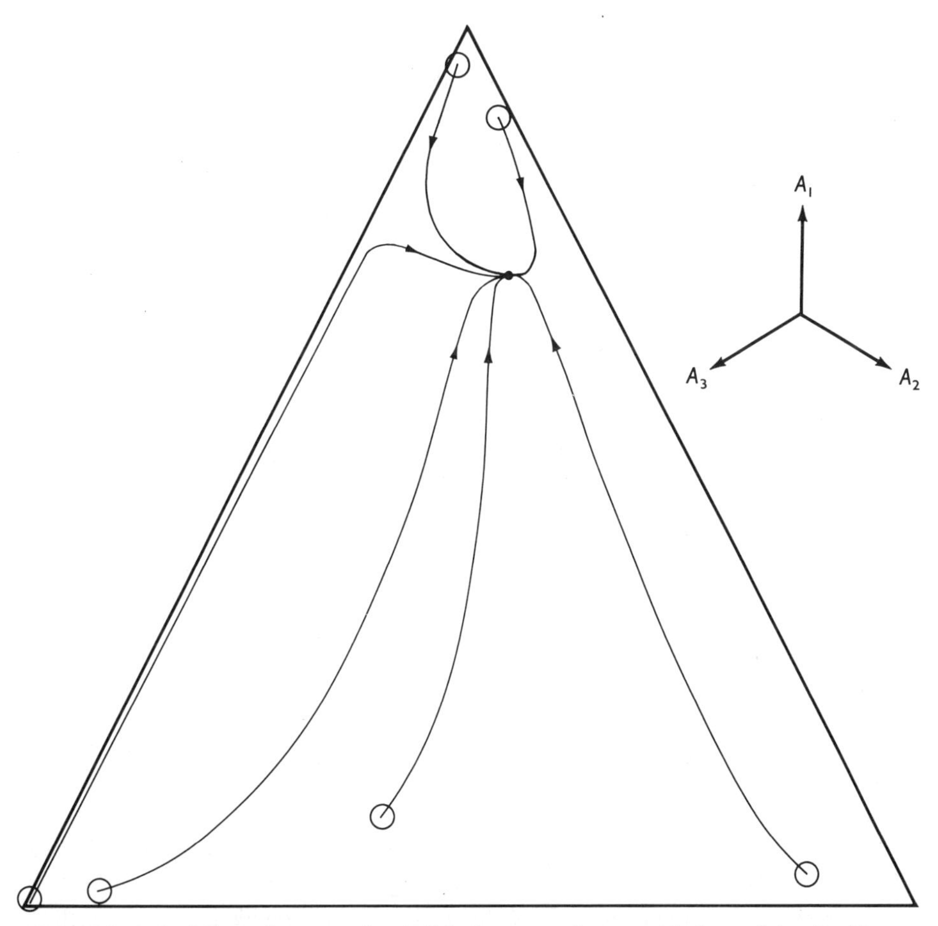

FIGURE 4-8. Still another example of allele-frequency changes with three alleles. In this case $w_{11} = 1.01$, $w_{22} = 0.99$, $w_{33} = 0.98$, $w_{12} = 1.02$, $w_{13} = 1.02$, and $w_{23} = 1.00$. A stable equilibrium is reached at $p_1 = 0.727$, $p_2 = 0.182$, and $p_3 = 0.091$. This example shows that all heterozygotes do not have to be more fit than any homozygote for a stable equilibrium to exist.

The rules for more than three alleles are obviously complicated. However, two simple statements apply to all situations in which the w's are constants:

1. **There is at most one stable equilibrium for two or more alleles. If a stable equilibrium exists, it is globally stable; that is, it will be reached from any starting point (provided, of course, that all three alleles exist in the population).**
2. **If a stable equilibrium exists, the mean fitness of the population exceeds that of any homozygote. If such a fit homozygote were to exist, it would be fixed in the population.**

Multiple-allele polymorphisms are often found. Examples are found in the ABO blood groups, the Rh factors, and the histocompatibility loci. Some of these polymorphisms involve several closely linked loci, but the amount of recombination is so small that the dynamics are essentially those of multiple alleles. Likewise there are many examples in which several alleles are maintained at isozyme loci. For example, the esterases often have multiple alleles at high frequencies in many animal populations.

Are these polymorphisms maintained by the selective forces that I have discussed? After their first discovery, isozyme polymorphisms were generally regarded as being maintained by a balancing selection that involved fitter heterozygotes. However, as the statistical analysis became more sophisticated, the hypothesis became weaker. Furthermore haploid organisms are just as polymorphic as diploids are. Therefore the overdominance hypothesis seems unlikely to be the major explanation, although it may account for some cases. One situation in which overdominance is almost certain to apply is inversion polymorphisms in Drosophila. In this case the overdominance is at the chromosomal level, not at the level of the individual gene loci.

4-9 FREQUENCY-DEPENDENT SELECTION

If the overdominance hypothesis is losing ground, as I indicated in the previous section, what are the selective alternatives? The two likely candidates are (1) frequency-dependent selection and (2) heterogeneous environment. Let us consider an example of frequency-dependent selection.

As mentioned previously, a natural definition of absence of dominance is that the heterozygote is the geometric mean, rather than the arithmetic mean, of the two homozygotes. This definition often leads to an algebraic simplification, which itself is perhaps sufficient justification for its use (see Problem 13 at the end of this chapter).

Table 4-8 gives a model in which the fitness is a function not only of the s's but also

TABLE 4-8

A model of frequency-dependent selection at a locus with three alleles. Each heterozygote has a fitness that is the geometric mean of the two relevant homozygotes.

GENOTYPE	A_1A_1	A_2A_2	A_3A_3
Fitness	$(1 + s_1p_1)^2$	$(1 + s_2p_2)^2$	$(1 + s_3p_3)^2$
Frequency	p_1^2	p_2^2	p_3^2
GENOTYPE	A_1A_2	A_1A_3	A_2A_3
Fitness	$(1 + s_1p_1)(1 + s_2p_2)$	$(1 + s_1p_1)(1 + s_3p_3)$	$(1 + s_2p_2)(1 + s_3p_3)$
Frequency	$2p_1p_2$	$2p_1p_3$	$2p_2p_3$

of the allele frequencies. Notice that the fitness of each heterozygote is the geometric mean of the two corresponding homozygotes.

We derive the recurrence equations in the now familiar way, leading to

$$\begin{aligned}\overline{w}p_1' = {} & p_1^2(1 + s_1p_1)^2 + p_1p_2(1 + s_1p_1)(1 + s_2p_2) \\ & + p_1p_3(1 + s_1p_1)(1 + s_3p_3),\end{aligned} \tag{4-36}$$

with corresponding equations for p_2 and p_3. The mean fitness is

$$\overline{w} = [p_1(1 + s_1p_1) + p_2(1 + s_2p_2) + p_3(1 + s_3p_3)]^2 = x^2. \tag{4-37}$$

The frequency-change equations simplify neatly to

$$p_i' = \frac{p_i(1 + s_ip_i)}{x}. \tag{4-38}$$

At equilibrium, $p_1' = p_i$; so

$$x - 1 = s_1\hat{p}_1 = s_2\hat{p}_2 = s_3\hat{p}_3.$$

Since the three allele frequencies must add up to 1,

$$\hat{p}_i = \frac{1}{s_iS} \quad \text{and} \quad S = \sum \frac{1}{s_i}. \tag{4-39}$$

Equation 4-39 gives us simple formulae for the equilibrium frequencies, but it does not tell us whether the equilibrium is stable or not. The equilibrium turns out to be stable if the s's are negative and unstable if they are positive. You can verify this rule by choosing allele frequencies close to the equilibrium and observing whether they move toward or away from the equilibrium.

The preceding conclusion is interesting and important. If s is negative, the rarer an allele is, the more favorable it is. Rare alleles tend to increase, which leads to a stable equilibrium. On the other hand, if each allele becomes more favorable as it becomes more frequent, there is no stable equilibrium. Rare alleles tend to become more rare, which means that even if the greater frequency of a rare allele would be beneficial to its bearer and to the population, it cannot increase. We shall return to this topic when we consider the evolution of cooperative behavior in Chapter 7.

In addition to heterozygote advantage, selection for rarity also leads to a selectively maintained polymorphism. The self-sterility alleles mentioned in Chapter 2 are an extreme example. We can easily see how rare alleles might be selectively advantageous in a very general way. The more similar two individuals are, the more they compete for the same resources. ("Your worst enemy is your brother.") A rare type would have the advantage of competing mostly with others who have different requirements. This advantage is certainly important in interspecies competition, but its role in individual competition or in maintaining variability is unknown.

I shall not discuss heterogeneous environments quantitatively. If several ecological niches are available and different genotypes are adapted to different niches, then the fitness will be highest when each genotype is in the right niche. Suppose that individuals are able to seek out their appropriate niches. A rare genotype is more likely to find a niche that is not completely full than a common genotype is. Thus multiple niches have selective properties similar to the one discussed in the preceding paragraph, in which rare alleles are at an advantage. For this reason a heterogeneous environment can promote polymorphism.

The literature contains much discussion of the role of environmental heterogeneity in preserving genetic variability. Once again, Darwin noticed it first. In one of his letters he stated that "wide ranging, much diffused, and common species vary most." The evidence is inconclusive as to the relative importance of balancing selection (perhaps involving interactions among several loci), mutation-selection balance, frequency-dependent selection, heterogeneous environment, or balance between neutral mutations and random drift in maintaining variability in natural populations.

4-10 SELECTION AT TWO LOCI

We have seen that with multiple alleles at a single locus, selection can be complicated, with sometimes unexpected and interesting results. Complicated selection patterns are even more likely when more than one locus is involved, especially with linkage. Multilocus theory is an active field of research, and I shall only touch the surface here. I shall concentrate on two situations that have attracted considerable attention among evolutionists.

In Chapter 1 we saw that in the absence of selection, linkage disequilibrium is reduced each generation by a fraction r, the amount of recombination between the two relevant loci. Therefore unless selection for certain linked combinations is intense or linkage very tight, most pairs of loci will be close to linkage equilibrium. Very roughly, unless the fitness of chromosomes in one linkage phase exceeds that of chromosomes in the other phase by an amount larger than the amount of recombination between them, the loci will be close to linkage equilibrium. I shall assume linkage equilibrium throughout this section.

Table 4-9 gives two models representing selection at two independent loci. We assume that the A and B alleles are completely dominant to a and b. We shall first consider model 1, in which the fitnesses are multiplicative.

TABLE 4-9

Selection at two independent loci. Two models are included: (1) fitnesses are multiplicative; (2) the A and B alleles are each disadvantageous alone, but together produce a beneficial effect. A and B are both assumed to be completely dominant. The constants s_A, s_B, and s_{AB} are all positive.

		FITNESS	
GENOTYPE	FREQUENCY	MODEL 1	MODEL 2
$A-\ B-$	$(1-p^2)(1-r^2)$	$(1+s_A)(1+s_B)$	$1+s_{AB}$
$A-\ bb$	$(1-p^2)r^2$	$1+s_A$	$1-s_A$
$aa\ B-$	$p^2(1-r^2)$	$1+s_B$	$1-s_B$
$aa\ bb$	p^2r^2	1	1

Multiplicative Fitnesses. Note first that we can write the mean fitness as the product of two terms:

$$\bar{w} = [p^2 + (1-p^2)(1+s_A)][r^2 + (1-r^2)(1+s_B)]. \tag{4-40}$$

Then we write the usual recurrence equation for q $(=1-p)$, which is

$$\begin{aligned} q' &= \frac{q[(1-r^2)(1+s_A)(1+s_B) + r^2(1+s_A)]}{\bar{w}} \\ &= \frac{q(1+s_A)}{1+s_A-p^2 s_A} \end{aligned} \tag{4-41a}$$

Notice that this equation does not involve either the frequencies or the fitness parameters associated with the other locus. Likewise the recurrence equation for s $(=1-r)$ is

$$s' = \frac{s\,(1+s_B)}{1+s_B-r^2 s_B} \tag{4-41b}$$

which also does not involve quantities associated with the other locus.

We reach the important conclusion, actually valid for any degree of dominance:

If the fitnesses are multiplicative between loci, the loci behave independently and we can use the single-locus equations of the earlier sections to describe allele frequency changes. This rule is approximately true if the loci are additive, provided the fitness differences are small.

This conclusion justifies our use of single-locus formulae in many population genetics equations. These formulae can only be approximations, of course, but if the loci act approximately independently on the phenotype, they are good approximations. Most classical population genetics theory assumes linkage equilibrium. Linkage is generally believed to have only second-order effects on most polygenic traits. An exact theory would have to take linkage into account, which much of the newer work in population genetics does — at the price of much more complicated algebraic expressions. However with multiplicative fitnesses there is no tendency to generate linkage disequilibrium.

Alleles That Are Disadvantageous Individually but Advantageous in Combination. Let us now assume the fitnesses of model 2 in Table 4-9. As you can see, a dominant A allele or a dominant B allele is disadvantageous by itself, but together they are beneficial. Many times in nature neither of two traits is advantageous by itself, yet the two coöperate to produce a beneficial effect. Usually the individual traits are not merely neutral, but disadvantageous, if for no other reason than that of wasting developmental resources that could be put to better use. Examples range all the way from intra-allelic complementation to lock-and-key mechanisms involved in mating and reproductive isolation in insects.

The recurrence equation for $q = 1 - p$ is now

$$q' = \frac{q[1 - r^2)(1 + s_{AB}) + r^2(1 - s_A)]}{\overline{w}}, \tag{4-42}$$

and the mean fitness is

$$\begin{aligned}\overline{w} = p^2r^2 + p^2(1 - r^2)(1 - s_B) + (1 - p^2)r^2(1 - s_A) \\ + (1 - p^2)(1 - r^2)(1 + s_{AB}).\end{aligned} \tag{4-43}$$

A useful device is to write the difference between successive generations, $\Delta q = q' - q$:

$$\Delta q = \frac{qp^2[s_B + s_{AB} - r^2(s_B + s_A + s_{AB})]}{\overline{w}}. \tag{4-44}$$

Setting $\Delta q = 0$, we find the equilibrium equation

$$r^2 = \frac{s_B + s_{AB}}{s_B + s_A + s_{AB}} \tag{4-45a}$$

and, by symmetry, its counterpart

$$p^2 = \frac{s_A + s_{AB}}{s_B + s_A + s_{AB}}. \qquad \text{4-45}b$$

This equilibrium is not stable. It is neither a maximum nor a minimum, but a saddle point. Figure 4-9 illustrates this situation. The two low points correspond to genotypes $A-$ bb and aa $B-$. The lower peak corresponds to aa bb and the higher

FIGURE 4-9. A fitness surface corresponding to the second model in Table 4-9. The two abscissas represent the frequencies of the A and B alleles, and the ordinate is the mean fitness of the population with these frequencies. A population starting with mostly aa bb genotyes cannot move to the higher fitness peak, since natural selection cannot move the population in the direction of decreased fitness.

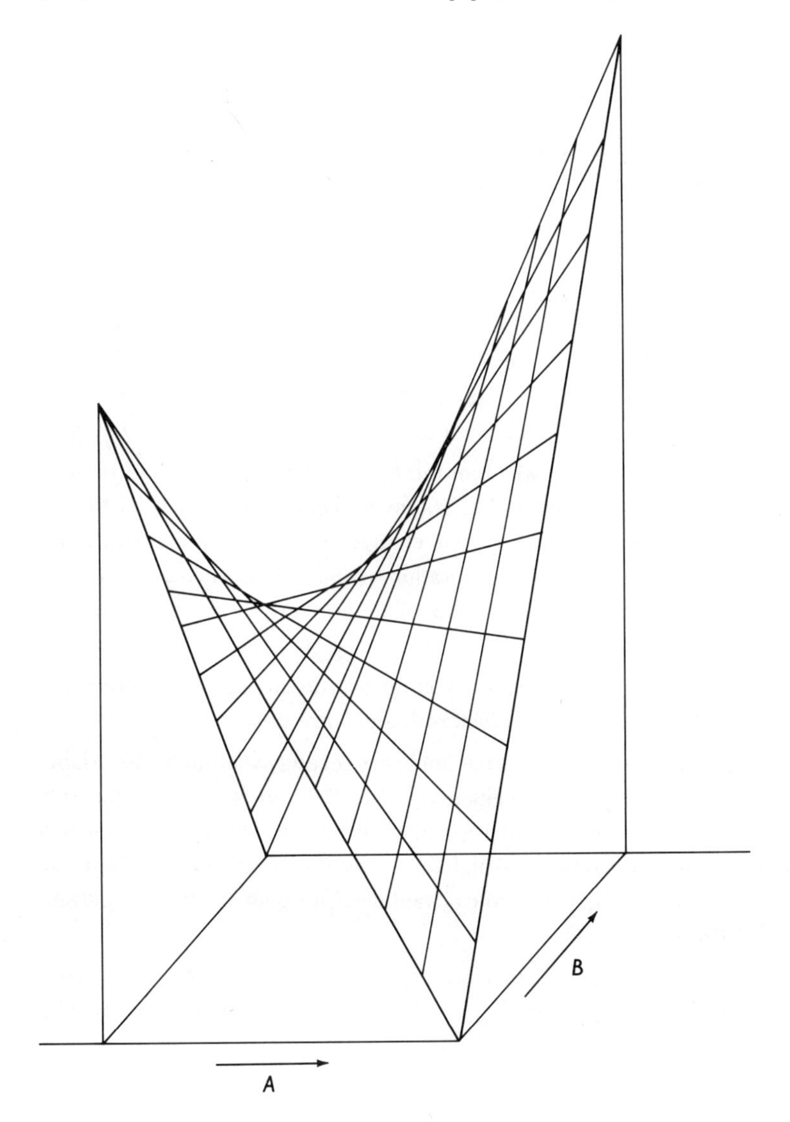

peak to $A-\ B-$. The most important conclusion is that if the population is $aa\ bb$, it cannot move to the higher fitness peak, since to do so involves going through points of lower fitness.

Another way to view this conclusion is to assume that the population is almost entirely $aa\ bb$ and that a new mutant A allele arises. From Equation 4-44, assuming that r is very nearly 1, we see that Δq is negative. The new allele cannot increase; it has no chance.

Is this inability to put together harmful alleles to get a beneficial interaction a serious impediment to evolution? Sewall Wright thinks it is, and this difficulty is the main motivation for his shifting-balance theory of evolution, which we shall consider in Chapter 7.

QUESTIONS AND PROBLEMS

1. Suppose that 100 individuals each of AA, AA', and $A'A'$ produce 120, 100, and 60 progeny, respectively. With the parameterization of Table 4-3, what are the values of s and h?
2. With selection coefficient s and dominance h, 5000 generations are required to change the frequency of an allele from 0.1 to 0.2. Answer one of the following questions, assuming s is small, and explain why the other questions are difficult. (a) How many generations are required if the selection coefficient is changed to $2s$? (b) How many generations are required if h is changed to $2h$? (c) How many generations are required to change from 0.2 to 0.3 if s and h are unchanged?
3. For a fixed absolute value of s, which of the following items are equal? (a) The number of generations required to change a favored dominant from 0.01 to 0.1. (b) The number of generations required to change an unfavored dominant from 0.1 to 0.01. (c) The number of generations required to change a favored recessive from 0.9 to 0.99. (d) The number of generations required to change an unfavored recessive from 0.99 to 0.9.
4. What is the equilibrium allele frequency at an overdominant locus if the three fitnesses are 1, $1 + hs$, and $1 - s$ (where h and s are positive)?
5. Cystic fibrosis is a recessive disease that until very recently was almost invariably lethal. Assume that $s = 1$. The incidence is 0.0004. (a) What mutation rate would be required to maintain such a frequency if the disease is completely recessive? (b) If this value seems too high for a typical mutation rate, perhaps the mutation is maintained by heterozygote advantage. Compute hs using the parameterization of Problem 4.
6. The frequency of sickle-cell anemia in one region of Africa is 1/144. Malaria has now been eradicated from this region. What will be the incidence of this anemia two generations from now?

7. Show that Equation 4-4 can be rewritten as $\Delta p = [p(1-p)/2\bar{w}](d\bar{w}/dp)$, where $p = p_1$ and $(1-p) = p_2$. Sewall Wright was responsible for this famous equation.

8. (a) Show that when there are only two alleles, Equation 4-9 can be written as $\Delta p = p_1 p_2(w_1 - w_2)/\bar{w}$. (b) Show that $\bar{w} = p_1 w_1 + p_2 w_2$.

9. Invent a pattern of viabilities so that the three genotypes remain in Hardy-Weinberg ratios after differential mortality.

10. Using the model of Table 4-6, what is the mean fitness when the population is at equilibrium?

11. A woman who has a brother with X-linked hemophilia asks you how likely it is that she is heterozygous for the allele. Assume that $s = 1$. (a) Assuming that the mutation rate is 10^{-5} in males and 10^{-6} in females, what is the probability that she is a carrier? (b) Do you need to know the mutation rate to answer this question or only the ratio of the male to female rate?

12. A haploid species has fitnesses in the ratio $1 : 1 - s$. Write equations analogous to Equations 4-5*b* and 4-21.

13. The fitness ratios in a diploid species are $1 : 1 - s : (1 - s)^2$. Again write equations analogous to Equations 4-5*b* and 4-21. Perhaps these equations suggest that the absence of dominance for fitness is best defined by making the heterozygote fitness the geometric mean of the fitness of the two homozygotes.

14. If the fitnesses of AA, AA', and $A'A'$ are 1, $1 - hs$, and $1 - s$, what range of values of h lead to (a) the ultimate loss of A'? (b) an unstable equilibrium? (c) a stable equilibrium?

15. Show that the fitnesses of Figure 4-8 lead to a stable equilibrium and compute the equilibrium allele frequencies.

16. In graphs such as those in Figures 4-6, 4-7 and 4-8, we assume that from any point in the triangle, the sum of the perpendicular distances to the three sides is a constant. Prove that this assumption is correct and that the distances add up to the altitude.

17. The rate of mutation to recessive lethals is 2.0×10^{-6} per gene per generation. The equilibrium frequency of lethal alleles in the population is 2.0×10^{-4}. (a) Are these measurements consistent with complete recessivity? (b) If not, compute h.

18. For the model of Table 4-6, what is the additive genetic variance when the population is at equilibrium?

CHAPTER 5

QUANTITATIVE TRAITS

Most of our knowledge about inheritance comes from the study of traits determined by either single allelic differences or a small number of differences. Part of Mendel's genius was his choice of clear-cut traits; it was his good luck that they turned out to be monogenic and not closely linked. The success of classical genetics in mutation analysis and chromosome mapping, as well as the spectacular achievements of molecular genetics, bear testimony to the power of genetic methodology using well-chosen marker loci.

Quantitative genetics, on the other hand, is concerned with differences of degree rather than kind. Phenotypes are classified not by the presence or absence of certain characteristics, for example, a gel band or a clear phenotype, such as singed bristles in Drosophila, but as values measured in centimeters, pounds, or bushels. Although the trait sometimes does not lend itself to direct measurements, individuals with the trait can be ranked. We can say that Joe is a better geneticist than John, but to say he is twice as good makes no sense.

Typically the inheritance of quantitative traits is complex. Many genes are usually involved — too many to be identified individually. Instead of trying to understand the effect of each gene, we treat them in the aggregate by statistical procedures. The most important characteristics for the animal or plant breeder, for human welfare, and for evolution usually depend on multiple factors. Darwin, with his usual perspicacity, wrote that such differences of degree "afford materials for natural selection to act on and accumulate, in the same manner as man accumulates in any given direction individual differences in his domestic productions."

5-1 THE NATURE OF QUANTITATIVE INHERITANCE

The genetics of quantitative traits is based on two fundamental premises:

1. The genes determining such traits are the same as those determining qualitative traits and differ only in that they do not produce a clear-cut phenotypic difference. We do not know whether such genes are mainly structural, regulatory, repeat sequences, small additions or deletions, of whatever kinds of DNA changes molecular biology has revealed or may reveal in the future.
2. The loci are cumulative in their effect. This does not mean that the genes are strictly additive but that if alleles *A* and *B* each increase size, the two together will produce a larger individual than either by itself.

The best evidence for these premises comes from the use of marker genes to dissect the quantitative trait and map the positions of the loci involved. A good example is DDT resistance in Drosophila. Figure 5-1 shows some representative data.

The resistant strain was selected in the laboratory by growing the flies in a cage that contained a population of a few thousand. The inner walls were painted with DDT, and the concentration was increased in successive generations as the flies became more resistant. After many generations flies from the resistant strain were mated to laboratory strains whose chromosomes could be identified by mutant markers. Figure 5-1 shows the survival rate of the different types. The mutant markers permitted the precise identification of each fly's genotype.

Clearly each of the chromosomes from the resistant strain contributes to the overall resistance. At least as many loci as chromosomes must be involved — actually there are more. A finer analysis identifying crossover products showed that several resistance-related loci are on each chromosome. At least half a dozen loci affect this trait, and possibly many more, since there is no clear upper limit.

Such studies, being laborious and unrewarding, are rare. One difficulty is that the marker genes themselves may affect the trait. Another is that environmental influences can be confused with genetic effects.

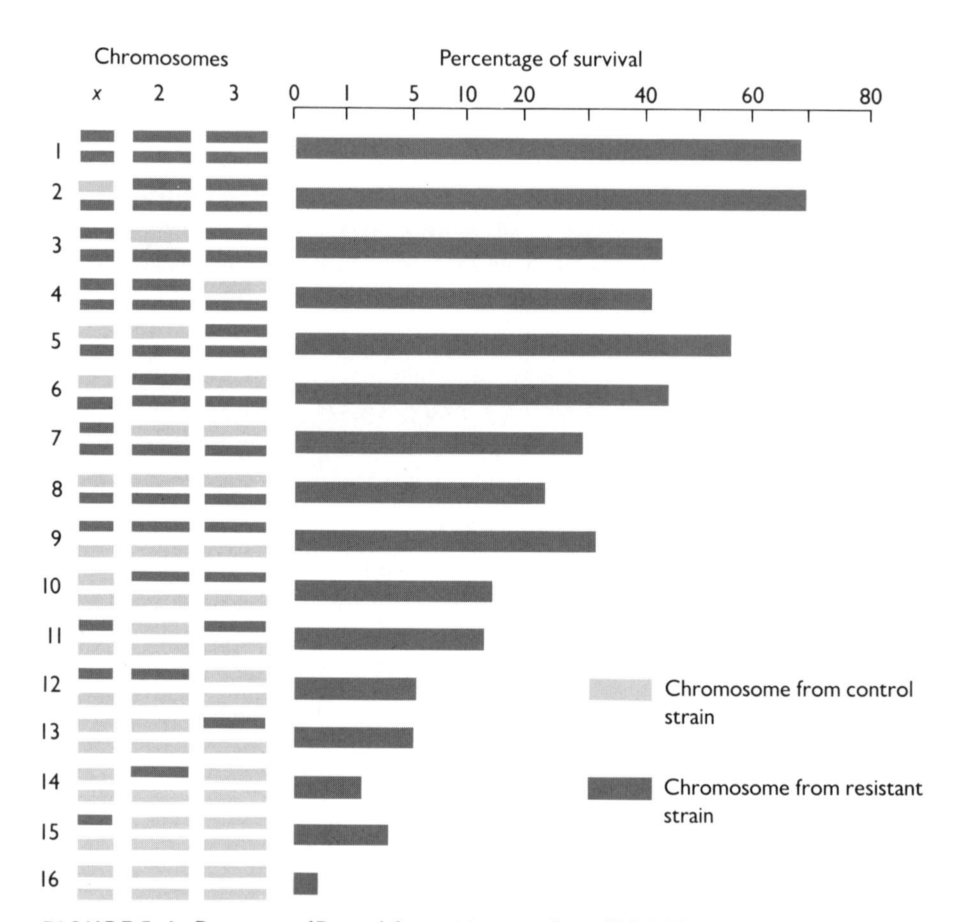

FIGURE 5-1. Percentage of Drosophila surviving a test dose of DDT. The Drosophila contain various combinations of chromosomes from a resistant and a normal strain. (Data from J. F. Crow. 1957. *Ann. Rev. Entom.* 2:227.)

The long-range aim of such research is to find what each allele is doing and how it interacts with each of the other alleles. Although molecular analysis will eventually clarify many examples, at present it is impractical and for many purposes unimportant.

At first glance we are surprised that polygenic effects are so nearly additive. When we consider the complexity of gene action, promoters, regulators, feedback loops, various complicated forms of epistasis, and the diversity of environmental effects, we might expect that gene combinations would be totally unpredictable. Yet most of the time an allele that increases a measurement in one environment or in combination with certain genes will do so in another environment or in another genotype. Although some striking, and fascinating, exceptions do occur, near additivity still holds, especially if the individual gene effects are small.

For natural selection to work, gene action must exhibit a certain predictability. If two genes that each increase a trait were to decrease the trait when combined, the effects of selection would be chaotic and ineffective. Also, there has to be a certain independence of body parts. For example, if weaker brains were produced every time nature selected for an improved liver function, evolution would be unable to improve simultaneously liver function and intelligence. I suspect that those species in which gene interaction is somewhat predictable are the survivors of the long struggle for existence. Perhaps that is why we find so much approximate additivity in nature.

I don't want to exaggerate this, however. *Perfect* additivity is rare, and much of quantitative theory is devoted to ways of taking account of nonindependence and nonadditivity, as you will see in later sections.

The Effectiveness of Long-Continued Selection. Almost *any* trait can be changed by selection over many generations. The insecticide resistance mentioned previously is a good example; incidentally, what were the genes that detoxify DDT doing before DDT was introduced?

One of the best examples of long-time selection is exhibited by a program begun at the Illinois Experiment Station in 1896 (before the rediscovery of Mendel's laws!). The experiment began with 163 corn ears with an oil content ranging from about 4 to 6 percent. By selecting in both directions, the experimenters soon developed strains that were well outside the original range of oil content, and the progress has continued ever since. Figure 5-2 shows the first 80 generations.

Comparison of the F_1 and F_2 variability from crosses between high and low strains suggests that these strains differ by at least 20 genes, many of which must have been segregating in the original population. Clearly the end has not been reached yet, for the strains are still diverging. The rate of change was fastest during the first 15 to 20 generations. Since then it has changed at about the same rate. Of course, selection produces a slower change going downward; you cannot go below 0.

Researchers have performed similar experiments for many traits in several species of animals and plants. Mice have been selected for color change. In one strain of dark-colored mice, one individual had a few white hairs on the forehead. After many generations of selection for more white, the average individual in the strain was 3/4 white and some individuals were entirely white. Rats have been selected for "intelligence" and "stupidity," as measured by maze-learning ability. Drosophila have been selected for positive and negative geotropism. Many other examples of various traits in many species could be cited.

Figure 5-3 gives some data on selection for large and small size in mice. As the left graph shows, the average weight in the large strain is almost double that of the small strain after some 21 generations.

The graph clearly shows that the rate of progress was more rapid during the early generations of selection that in the later generations. This result is typical; we saw it in

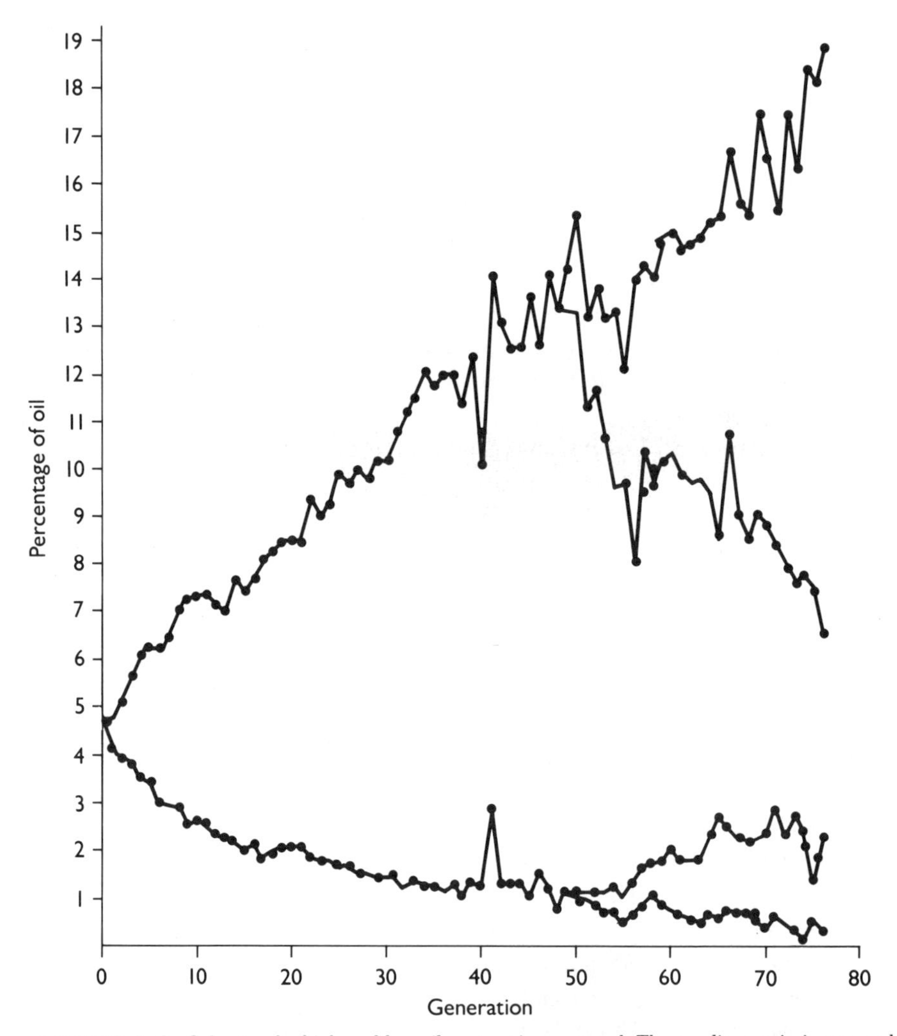

FIGURE 5-2. Selection for high and low oil content in corn seed. The two lines pointing toward the center indicate the results of selection in the reverse direction started in generation 48. (Data from J. W. Dudley. 1977. *Proc. Int. Cong. Quant. Genet.;* E. Pollak, O. Kempthorne, and T. B. Bailey, eds., pages 459 ff.)

the Illinois corn experiment. The explanations are diverse. Some of the segregating loci may have become homozygous, so that they cannot be changed without new mutations. Often as a strain departs from its original value, it has reduced viability and fertility. It is as if a mouse has an optimum size determined by its long history of natural selection, and any large departure from this size produces an animal whose machinery works less smoothly.

Another feature of the experiment is that selection produces a greater effect in the

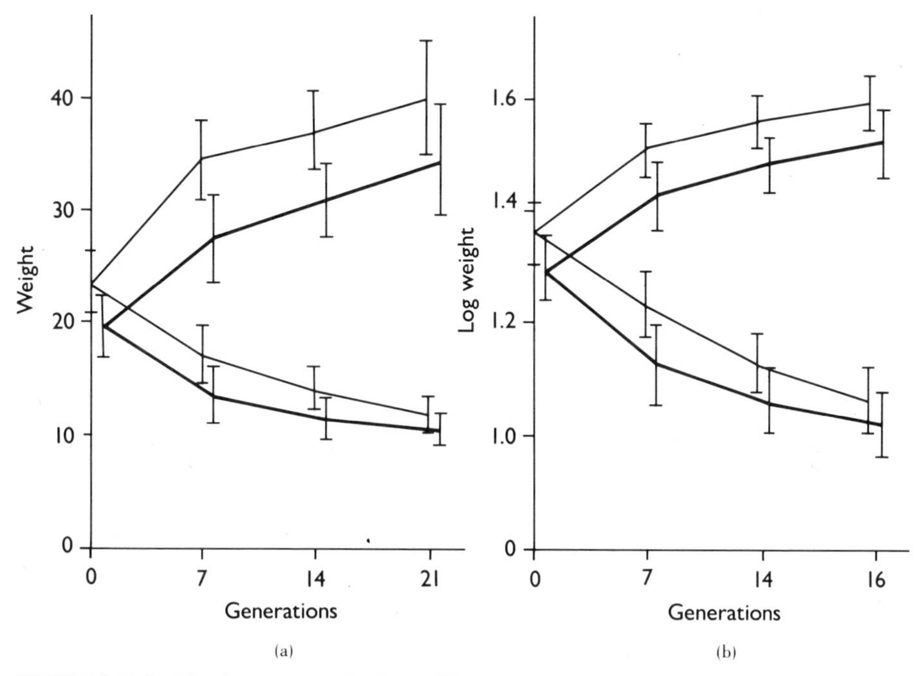

FIGURE 5-3. The change in weight during 21 generations of selection for large and small size in mice. Note that the change is more nearly equal in the two directions and that the standard deviations (indicated by vertical bars) are more similar when the scale is logarithmic (a) than when arithmetic (b). (Data from J. W. MacArthur. 1949. *Genetics* 34:194 ff. As analyzed by S. Wright. 1977. Volume III, page 204.)

upward direction than in the downward direction. This pattern was also true of the corn oil experiments. In the next section we shall consider how to simplify the analysis of such asymmetrical trends.

Multiplicative Gene Action. There is good evidence that size genes, and perhaps many others, act multiplicatively rather than additively. It is as if a particular allele, rather than adding 1 millimeter to the length of a rat, adds 1 percent of the existing length. We might expect the same gene to add more millimeters to an elephant than to a mouse. For small differences we cannot distinguish between additive and multiplicative effects, but when the size differences are large the effects may be quite dramatic.

Table 5-1 gives data on two parental strains of tomatoes and their F_1 hybrid. We can see that the F_1 values are much nearer to the geometric means than to the arithmetic means. This pattern, by itself, does not prove that the genes act multiplicatively; for example, small-size alleles could tend to be dominant. Yet the consistency of these results suggests multiplicative effects.

We can obtain better evidence by looking at the F_1 and F_2 generations from two

TABLE 5-1

Mean weight of fruit in grams from crosses between tomato strains with widely divergent fruit sizes. P_L and P_S stand for the mean weight values of the large and small parent strains.

					MEAN P	
LARGER PARENT	SMALLER PARENT	P_L	P_S	F_1	ARITHMETIC	GEOMETRIC
Large Pear	Red Currant	54.1	1.1	7.4	27.6	7.7
Putnam's Forked	Red Currant	57.0	1.1	7.1	29.0	7.9
Tangerine	Red Currant	173.6	1.1	8.3	87.3	13.8
Devon Surprise	Burbank President	58.0	5.1	23.0	31.5	17.2
Honor Bright	Yellow Pear	150.0	2.4	47.5	81.2	43.1

Data from J. W. MacArthur and L. Butler. 1938. *Genetics* 23:253.

homozygous parental strains. If the genes act additively, the F_2 mean should be the same as $P_1/4 + F_1/2 + P_2/4$, where P_1 and P_2 are the means of the parental strains and F_1 is the mean of the F_1 hybrids. This rule follows from simple Mendelian ratios for a single locus. If the loci are additive, the total effect of all loci should produce values that have the same relationship as the individual loci. In the tomatoes such experiments showed very poor agreement with this expectation, which indicates that the genes are not acting additively.

The mouse data in Figure 5-3 also suggest multiplicative gene action. The figure on the right plots the logs of the weights as the ordinate. Multiplying numbers is equivalent to adding their logs; so we can convert multiplicative to additive effects by plotting logs. When the logs are used, we can see that the curves are more nearly symmetrical around the starting point. Additional evidence comes from the fact that the variability, as indicated by the vertical lines, is more nearly the same in the upward and downward selection when the logs are plotted. The additivity tests with the tomato data were also much improved when log weights were used instead of the actual weights.

Since much of quantitative genetics depends on analysis of variability, it is important that the total variability be roughly constant when the mean changes. In practice, geneticists try to achieve this condition through transformations of the original data, as I did in the right part of Figure 5-3 by using the logs of the original data. The transformation used is often the log function but it can be square roots or a more complex function — anything that makes the variability constant over the range of interest is acceptable. The geneticist then performs the appropriate calculations with the transformed data and, when finished, uses the reverse transformation (for example, antilogs) to convert the results back to the original units. In addition to stabilizing the variability, the transformation of data usually makes the distribution more normal (see Appendix 3) and the response to selection more symmetrical.

Table 5-2 gives some data on height and weight of male university students. If we plot the data, they produce a bell-shaped curve that is nearly symmetrical for height but is skewed to the high side for weight. A simple, but effective, way of testing whether or not the data are normally distributed is to plot the cumulative frequency on graph paper on which the ordinate is scaled according to the integral of the curve of normal distribution (see Figure 5-4). Appendix 3 gives the equation for this curve.

If the data are normally distributed, the plotted points will fall on a straight line. If the data are skewed, the points will lie on a curve. Since the human eye can easily detect departures from a straight line, this is a sensitive test of normality. Figure 5-4 shows graphically the data from Table 5-2. Notice that the cumulative points for height lie close to a straight line, but those for weight are clearly curved. The third plot is for the logs of the weights. These values, as you can see, fall on a straight line, arguing that the factors influencing weight tend to be multiplicative rather than additive.

TABLE 5-2

Height and weight of 1000 male Harvard students.

HEIGHT (cm)	FREQUENCY	CUMULATIVE FREQUENCY	WEIGHT (kg)	FREQUENCY	CUMULATIVE FREQUENCY
156	4	4	45	1	1
159	8	12	48	5	6
162	26	38	51	22	28
165	53	91	54	58	86
168	89	180	57	85	171
171	146	326	60	142	313
174	188	514	63	154	467
177	181	695	66	151	618
180	125	820	69	138	756
183	92	912	72	100	856
186	60	972	75	60	916
189	22	994	78	34	950
192	4	998	81	25	975
195	1	999	84	12	987
198	1	1000	87	7	994
			90	2	996
			93	2	998
			96	0	998
			99	0	998
			102	0	998
			105	2	1000

Data from Wright. 1968. Volume I.

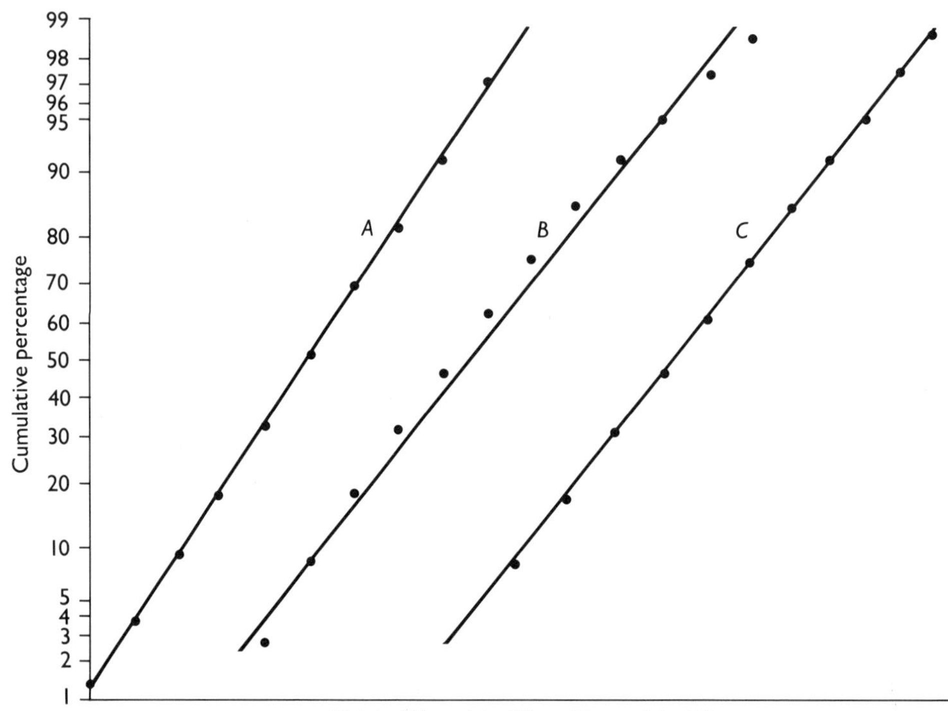

FIGURE 5-4. A plot of the cumulative frequency for height *(A)*, weight *(B)*, and log weight *(C)* of male Harvard students. The ordinate is scaled so that the cumulative normal distribution is a straight line. By this graphical criterion height is normally distributed, but weight is not. However, log weight is normal.

Henceforth we shall assume that the data behave, or have been transformed to behave, as if the effects were approximately additive and the distribution normal. Often, especially if the differences being studied are small relative to the means, the transformation has little effect and is omitted.

5-2 REGRESSION AND RESPONSE TO SELECTION

In the nineteenth century Francis Galton, Charles Darwin's cousin, coined the expression **filial regression,** or simply **regression,** to describe the tendency for the progeny of extreme parents to deviate from the population average by a smaller amount than their parents did. The explanation is complex, depending on dominance, epistasis, and effects of the environment; I shall try to untangle some of the causes later in the chapter. Much of quantitative genetics is concerned with understanding regression and predicting the results of selection when regression occurs. The word *regres-*

sion has become part of the standard vocabulary of statistics and is now used in a more general way, as we shall see in Section 5-3.

Figure 5-5 illustrates the phenomenon of regression. To be concrete, let us think of size. The graph shows the frequency of different-sized animals in the population. The abscissa is size, and the ordinate represents the frequency of animals of this size. The distribution is roughly symmetrical, either naturally or as a result of transformation.

The heavy arrows indicate the average size of the progeny. Progeny of large parents tend to be smaller than their parents; progeny of small parents tend to be larger. If this were a complete description, we would expect the population to become less variable each generation, which of course does not happen. As indicated by the other arrows, the individual progeny vary around the average of their group, so that the population distribution remains the same from generation to generation (in the absence of selection and ignoring random fluctuations). The spread outward from the large number of parents near the center compensates for the inward tendency of the progeny of parents high and low on the size scale.

Geneticists use the word **heritability** to measure the effects of regression. Since heritability is also defined in another way, I shall call this term **heritability in the narrow sense,** designated H_N. The broad-sense use of the word will be discussed later in the chapter. If the progeny deviate from the average by 1/4 as much as the parents did, the heritability is 1/4. The parents can deviate in either direction, but the

FIGURE 5-5. Regression toward the mean in the progeny of extreme parents.

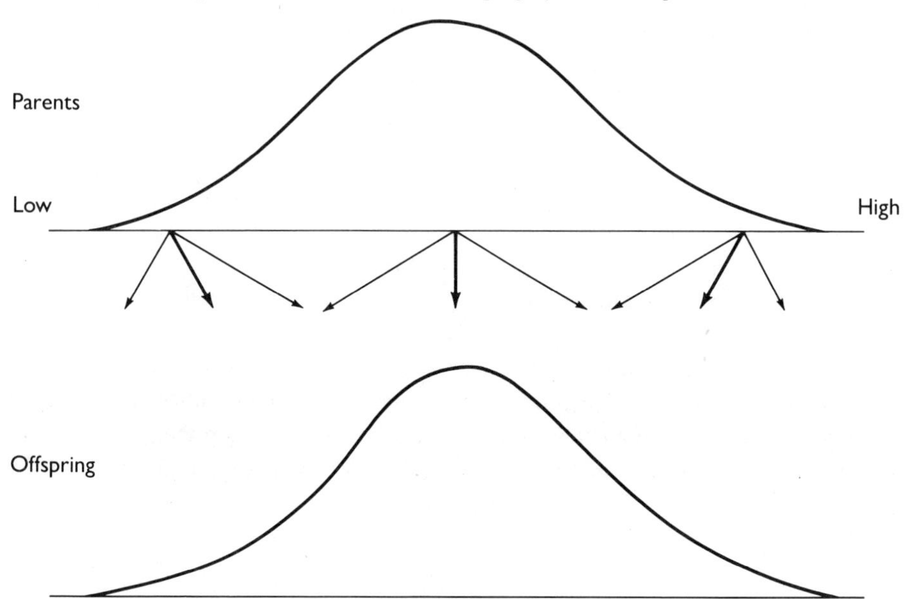

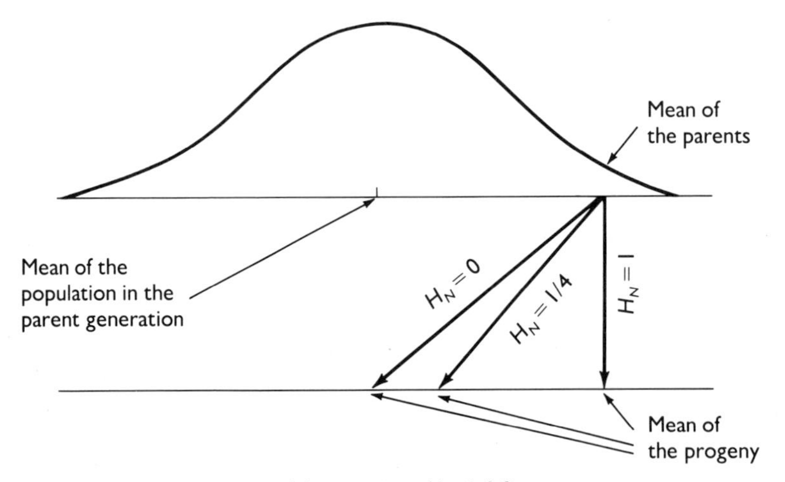

FIGURE 5-6. Illustration of the meaning of heritability.

progeny deviate (on the average) in the same direction as the parents did. Thus we can use heritability for selection either upward or downward. Figure 5-6 illustrates the meaning of heritability.

Figure 5-7 illustrates two generations of selection for a trait with heritability 1/3. The breeder **truncates** the population by using, as parents, only those individuals above a certain size. The mean of the parents is somewhere to the right of the truncation point. If many genes are involved and the trait is partly determined by the environment, as it almost always is, these effects are scrambled in the next generation and the progeny of the selected parents — despite the very asymmetrical distribution of the parents — are again approximately normally distributed.

In Figure 5-7 we assume that the heritability is 1/3. The average of the progeny deviates from the population mean by 1/3 as much as the parents did. Then a selected portion of the progeny becomes the parents of the next generation, and again the mean is shifted by 1/3 the amount that the new parents deviated from the mean of *their* generation. The population increases in size each generation but at only 1/3 the rate a heritability of 1 would produce. Selection with heritability 1/3 is like climbing a slippery ladder, in which one reaches up three rungs but always slides back two. The climber gets to the top eventually but at only 1/3 the rate with a nonslippery ladder.

This principle is usually stated as a standard formula, which you can understand by referring to the figures. The principle states that the difference between the progeny mean and the population mean is equal to the difference between the parental mean and the population mean multiplied by the heritability. If the deviations are negative, the rule still works. The predicted value of the offspring is

$$\text{Offspring} = \text{population mean} + H_N(\text{parent mean} - \text{population mean}). \qquad 5\text{-}1$$

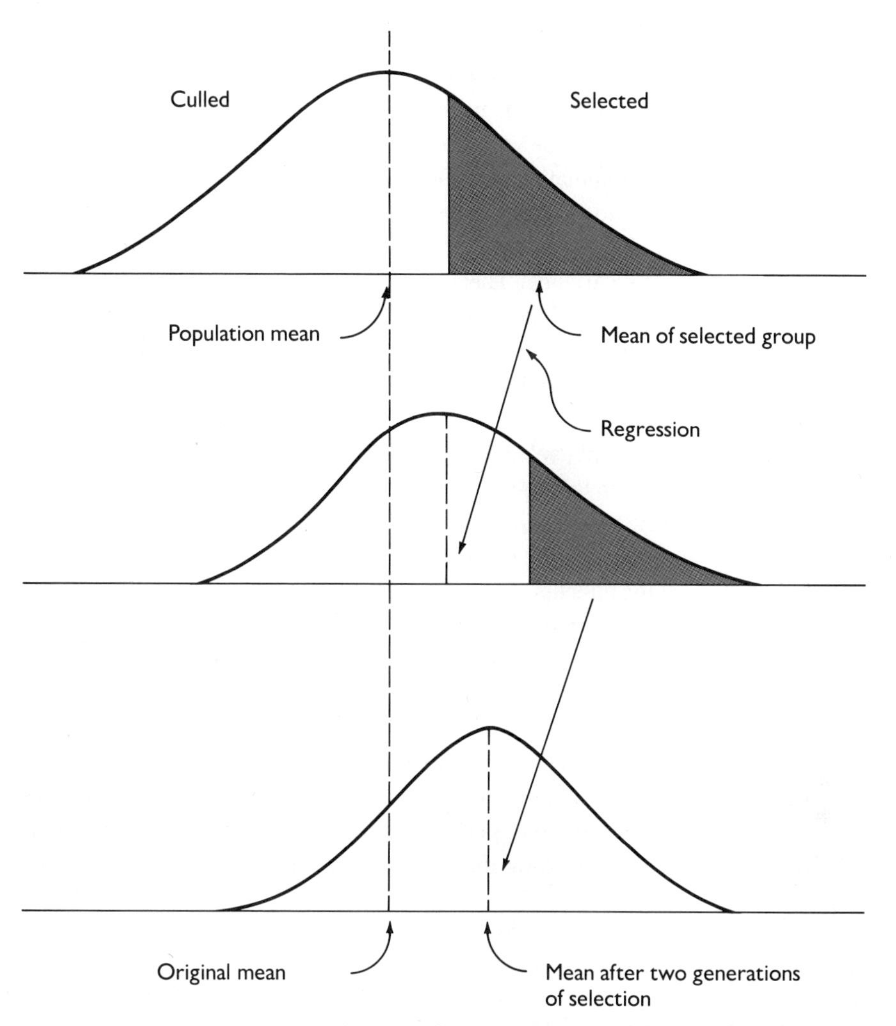

FIGURE 5-7. Two generations of selection for a trait with a heritability of 1/3. Each generation the progeny mean moves 1/3 of the distance from the population mean to the mean of the parents.

A practical problem arises when there are systematic mean differences in the two sexes. One way that breeders resolve this problem is to standardize all values to those for, say, females by subtracting from each male value the difference between the female and male means.

Why Doesn't Selection Reduce the Variance? If the selected parents are only a fraction of their generation, as Figure 5-7 implies, they have a smaller variance than the population they came from. Yet I have portrayed the progeny generation as having the same variance as the parent generation. How can this be true?

One reason is that part of the variability is caused by the environment, and this part remains in the next generation. As for the genetic component, if many loci are involved, individual allele frequencies do not change very much. The contribution that an allele of frequency p makes to the variance of the trait is proportional to $p(1 - p)$. For an explanation of this mathematical relationship, see the discussion of the binomial distribution in Appendix 3. Now if p is less than 1/2, its contribution to the variance will increase as its frequency increases. If p is greater than 1/2, increasing its frequency will decrease the variance. Therefore if alleles are segregating at a large number of loci, selection will be increasing the variance contribution of some and decreasing that of others; the net change will be small. Hence with multiple genes selection can continue many generations with a substantial change in the mean and little change in the variance.

If the number of loci is small, selection may carry some of them to homozygosity, which causes the population to lose variability. The importance of this effect depends both on the number of loci involved and on the size of the population; small populations may become homozygous through the combined effects of selection and random drift.

The whole picture of the effect of long-continued selection on the genetic variance is not clear and may differ from case to case. A full understanding would require knowledge of how many loci are involved, how intense the selection is, and the mutation rate.

I suspect that the main barrier to extreme changes by selection, at least in large populations, is not the loss of genetic variability but antagonistic selection forces. For example, too large a size may lead to weakness or infertility, so that artificial selection and natural selection are working in opposite directions. Not surprisingly the rate of progress by selection eventually slows down, as the examples early in this chapter illustrate.

5-3 SOME RELEVANT STATISTICAL MEASURES

Appendix 2 gives several statistical formulae (see Table A-1) and discusses the computational procedures. For convenience I have repeated the most important formulae in Table 5-3.

The formulae for variance and covariance have N in the denominator. If you are doing actual computations from data, divide by $N - 1$ rather than by N. As explained in the appendix, if we make the measurements on a sample and want to use these measurements to estimate the values in the population from which the sample was drawn, we can calculate an unbiased estimate by dividing by $N - 1$.

Variance. The variance is the average of the squared deviations from the mean. Its square root is called the standard deviation. The variance is difficult to interpret

TABLE 5-3

The principal statistical measures used in quantitative genetics. In the weighted column, f is the frequency or the weighting factor.

	UNWEIGHTED	WEIGHTED
Mean	$\bar{X} = \dfrac{\Sigma X}{N}$	$\dfrac{\Sigma fX}{\Sigma f}$
Variance	$V_X = \dfrac{\Sigma[(X - \bar{X})^2]}{N}$	$\dfrac{\Sigma[f(X - \bar{X})^2]}{\Sigma f}$
Covariance	$C_{XY} = \dfrac{\Sigma[(X - \bar{X})(Y - \bar{Y})]}{N}$	$\dfrac{\Sigma[f(X - \bar{X})(Y - \bar{Y})]}{\Sigma f}$
Standard deviation	$s_X = \sqrt{V_X}$	
Correlation	$r_{XY} = \dfrac{C_{XY}}{s_X s_Y}$	
Regression	$b_{XY} = \dfrac{C_{XY}}{V_X}$	
Variance of mean	$V_{\bar{X}} = \dfrac{V_X}{N}$	

because it is measured in squared units; that is, if we are measuring height in inches, the variance is in squared inches. One way to eliminate this dimensional inconsistency is to take its square root and use the standard deviation. However, by so doing we lose a more important property—additivity. Therefore variances are better suited than standard deviations for quantitative genetics.

Variances are additive in the following sense: if a total measurement is the sum of a number of components and if these components are independent, the variance of the sum of the measurements is the sum of the individual variances. For an explanation and derivation see Equation A-8 and the corresponding text in the appendix. Furthermore, as I explained in Chapter 4, selection changes a population in proportion to its genetic variance, not its standard deviation or any other measure of variability.

Covariance. As is clear from its definition in Table 5-3, the covariance is a measure of the similarity of two measurements, X and Y. If each X is the same as its corresponding Y, then the covariance is the same as the variance. If X and Y are independent, the covariance will be 0, because the deviations in opposite directions that lead to minus signs will balance those in the same direction that yield plus signs. We shall use covariances to measure similarity between relatives. Covariances, like variances, are scaled in the products, or squares, of the original units and are additive.

Correlation. We often wish to rescale the covariance to provide a dimensionless quantity that lies in the range -1 to $+1$. We obtain this quantity, the **coefficient of correlation,** by dividing the covariance by the product of the two standard deviations. If Y and X are perfectly correlated (each Y is a linear function of the corresponding X value), then the correlation is 1; otherwise it is less than 1. If Y and X are independent, the correlation is 0. If a large Y tends to go with a small X, then the correlation is negative.

Regression. The regression of Y on X, b_{YX}, is the slope of the line that best fits the data points. See Figure A-2 in the appendix for an illustration. Technically the line of best fit minimizes the sum of the squares of the vertical deviations of the points from the line. You can find a derivation of this **least-squares** procedure in any statistics textbook or in C&K, pages 487 ff. A least-squares line will always pass through the means of the X's and Y's.

We ordinarily write the equation of the least-squares line in the form

$$Y - \bar{Y} = b_{YX}(X - \bar{X}). \tag{5-2}$$

In words, the amount by which a predicted Y value deviates from its mean is equal to b_{YX} times the amount by which the corresponding X value deviates from its mean.

If we compare Equations 5-1 and 5-2, we see that they are essentially the same. If we identify the predicted offspring with Y, the mean of the parents with X, and the population mean with the mean of X and Y, then b_{YX} is equivalent to H_N. The explanation for this correspondence is complicated by Mendelian segregation, which we will discuss in Section 5-5.

Covariance of a Part with the Whole. Suppose that a measurement, Z, is the sum of two components, X and Y, and that these two components are independent. For example, X might be the genotypic component of corn yield and Y the environmental component. We can write the covariance as

$$\begin{aligned}
\mathrm{Cov}_{XZ} &= \frac{\Sigma[(X - \bar{X})(Z - \bar{Z})]}{N} = \frac{\Sigma[(X - \bar{X})(X + Y - \overline{X + Y})]}{N} \\
&= \frac{\Sigma\{(X - \bar{X})[(X - \bar{X}) + (Y - \bar{Y})]\}}{N} \\
&= \frac{\Sigma(X - X)^2}{N} + \frac{\Sigma[(X - \bar{X})(Y - \bar{Y})]}{N}.
\end{aligned}$$

In the last expression, the first term is the variance of X and the second is the covariance of X and Y, which is 0 since X and Y are independent. This leads to

$$\mathrm{Cov}_{XZ} = V_X. \tag{5-3}$$

Notice that this relationship is the same whether we divide by N or $N - 1$.

We are now equipped with the necessary information to consider the allocation of phenotypic differences to genetic and environmental causes.

5-4 HERITABILITY IN THE BROAD SENSE

An individual that is larger than the average of the population may have more alleles that increase size or may have been reared in an environment favorable to larger size. Usually both influences are involved: part of the excess height is caused by the genotype and part by the environment. Figure 5-8 depicts this situation.

We designate the relevant measurements for a particular individual i as follows:

M = the mean value of the population

P_i = the phenotypic value of individual i

H_i = the genotypic (hereditary) value of individual i.

We obtain the value P_i directly by measuring the characteristic of the individual under consideration and the value M by averaging the measurements of all individuals in the population. H_i is more abstract and must be obtained indirectly. We can scale these values to more workable numbers by subtracting M. I shall use lowercase letters to designate the quantities that are scaled this way. P_i can as easily be below the population average as above it, in which case h_i and p_i are negative.

Throughout this section and the next two, we assume that the genetic and environmental effects are additive, that is, that there is no interaction. We also assume that the genetic and environmental factors are randomized, so that there is no covariance; that is, there is no tendency for individuals with a genotypic propensity for tallness to be in environments that are also conducive to tallness, or vice versa. Later, in Section 5-7, we shall consider these complications.

FIGURE 5-8. Genotypic and phenotypic values for a quantitative trait. Uppercase letters designate absolute values; lowercase letters denote deviations from the mean.

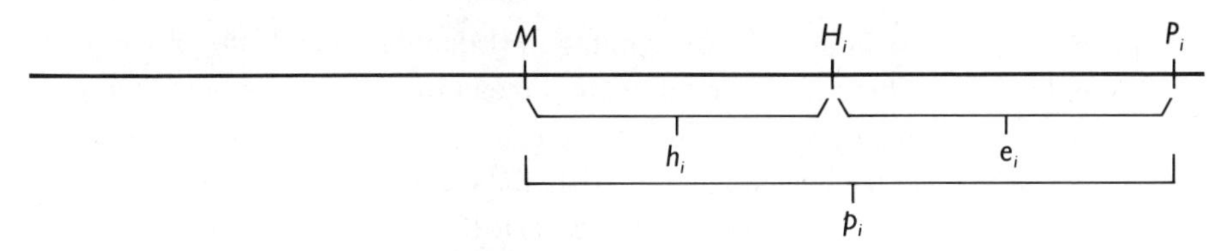

The phenotypic deviation, p_i, of individual i is the sum of the genotypic, h_i, and environmental, e_i, deviations:

$$p_i = h_i + e_i. \tag{5-4}$$

The mean values of both h and e, and therefore p, are 0.

We can also determine the variance of the measurements of all the individuals in the population or in a sample. Because of the additivity of variances when genotype and environment are independent,

$$V_P = V_H + V_E. \tag{5-5}$$

Heritability in the broad sense is defined as

$$H_B = \frac{V_H}{V_P}. \tag{5-6}$$

We can look at the broad sense heritability in another way. Since h is a part of p, we can substitute C_{HP} for V_H from Equation 5-3, leading to

$$H_B = \frac{C_{HP}}{V_P} = b_{HP}. \tag{5-7}$$

Interpretation of the Broad-Sense Heritability. Equations 5-6 and 5-7 illustrate the two interpretations of the broad-sense heritability: (1) as a ratio of genotypic to phenotypic variance and (2) as the regression of genotype on phenotype. Let us consider both of these interpretations.

1. Variance, or population, interpretation (Equation 5-6). The heritability is the proportion of the population variability, measured as the variance, that is attributable to genetic differences; the remaining proportion is caused by environmental influences. If all the individuals could be reared in identical environments, the phenotypic variance, V_P, would be reduced to V_H. Alternatively if the individuals were all of the same genotype, the variance would reduce to V_E.
2. Regression, or individual, interpretation (Equation 5-7). If the phenotype of individual i differs from the population mean by p_i, then the predicted genotypic deviation, h_i, is $H_N p_i$. For example, if the mean height of human males is 70 inches and the heritability is 0.8, then the genotypic deviation of a man 80 inches tall (10 inches above the average) is 8 inches. He can attribute 8 of the

10 inches by which he exceeds the population average to his genotype and the other 2 inches to his environment. If this man had been reared in an average environment, we would predict his height to be 78 inches. Of course, these interpretations are based on averages. Individual values differ widely.

Estimation of the Broad-Sense Heritability. The best way to estimate broad-sense heritability is to compare the variance of organisms having the same genotype with that of organisms having randomly varying genotypes, with both groups having been reared in the same randomized environment. We can easily make this type of comparison in species that can be propagated vegetatively or species in which inbred lines and their F_1 hybrids are available.

A guinea pig experiment by Sewall Wright some 60 years ago is still one of the best illustrations of broad-sense heritability. The trait being measured is the amount of white spotting. Wright devised a scale transformation that makes the variance independent of the amount of white. As shown in Table 5-4, the variance of the original randomly mated population, V_P, was 573 units. After many generations of sib mating, the average variance within an inbred line was 340. Since this variance is caused entirely by the environment, it estimates V_E. We can then obtain V_H by subtracting 340 from 573. The column on the right converts V_E and V_H into fractions of the total, V_P, giving a broad-sense heritability of 0.407. About 40 percent of the variance in the original population is genetic and 60 percent is environmental; the variance is 100 percent environmental in an inbred line. Wright tested the inbred lines by measuring the correlations between relatives within a line. These correlations were not significantly different from 0, as we would expect if the populations are essentially homozygous. In a homozygous population two relatives are no more alike than any two random individuals.

If we have several sets of identical twins reared in independent environments, we can measure H_B directly from their correlation. Within one set of twins, each individual has a genotypic contribution shared by the other and a separate environmental contribution. The phenotypic deviations of twins 1 and 2 are $p_1 = h + e_1$ and p_2

TABLE 5-4

Variance components for white spotting in guinea pigs. The variance is measured in arbitrary units on a scale transformed to make the variance independent of the mean value.

COMPONENT	VARIANCE	FRACTION OF TOTAL
V_H	233	0.407
V_E	340	0.593
V_P	573	1.000

$= h + e_2$. The covariance of p_1 and p_2 is $\Sigma(h + e_1)(h + e_2)/N$, which is equal to $\Sigma h^2/N$, since e_1, e_2, and h are all uncorrelated. Thus $C_{P_1P_2} = V_H$ and $H_N = V_H/V_P = C_{P_1P_2}/V_P$. Since the two twins have the same variance, the regression is equal to the correlation, and $H_B = \text{r}$.

We must allocate the two members of a twin set to random environments. For example, if we put several sets of identical cattle twins into the same field, the twins tend to stay together and therefore encounter similar environments. The environmental similarities resulting from this behavior might mistakenly be classified as genetic. The same problem arises with separated human twins. Are their environments as different as those of random persons in the population?

For example, the Chicago zoologist H. H. Newman studied 19 sets of one-egg twins reared in different homes and found that the correlation for height was 0.95 and for weight was 0.90. There is considerable doubt that the environments were really independent; they were probably more similar than if they were chosen at random. Thus the correlation is probably an overestimate of the heritability. Although the exact value of the correlation is questionable, it is apparently high, which argues that within the United States genetic differences are more important in determining variability in height than are environmental differences.

Notice that this type of analysis does not identify the actual environmental causes. In addition to the tangible variables such as temperature and nutrition, the environment includes events such as embryological accidents, which we do not ordinarily think of as environmental. It is a strength of heritability analysis that it can tell us how much environmental differences are contributing to the population variance even when the nature of these environmental factors is unknown. It is a weakness of such studies that they offer no guidance as to where to look for the causes of the environmental differences.

5-5 A SINGLE-LOCUS MODEL

In the previous section we considered variances, heritability, and regression in situations where genetically identical individuals can be studied in different environments. Such studies are applicable, for example, to identical twins, clonally propagated plants, and the F_1 progeny of inbred lines. We shall now consider the effects of the Mendelian gene-scramble that characterizes sexually reproducing, noninbred populations. What is transmitted from parents to offspring is not a genotype but a random sample of alleles, and the average properties of these alleles determine the characteristics of the offspring.

Ordinarily we do not know what is going on at individual loci. Therefore we resort to statistical measures of the aggregate effects of all the relevant loci. We measure these effects by such quantities as variances and covariances. But to understand the complicated effects of Mendelian segregation, we must first understand what happens at a single locus. We can then generalize this information to a large number of loci.

Genic and Genotypic Values. The general single-locus model is very similar to the model used in Chapter 4, particularly the one used to derive Equation 4-9. At that time we used w to represent the fitness. We shall now use Y to symbolize the quantitative trait of interest; for mnemonic assistance think of Y as standing for yield. Let Y_{11}, Y_{12}, and Y_{22} represent the mean values of individuals of the three genotypes, as shown in Table 5-5.

The values of the Y's are determined mainly by other loci and the environment; so the a_{ij}'s are small relative to $\bar{Y}$. As in Equation 4-10, we define the average effect of allele A_1 in analogy with the average excess of this allele on fitness, $w_1 - \bar{w}$. Therefore as Table 5-5 shows, $a_1 = p_{11}a_{11} + p_2 a_{12}$.

For our purpose the words *excess* and *effect* are synonymous. They are different in more general treatments but reduce to the same value when the population is in Hardy-Weinberg proportions and linkage equilibrium (see C&K, pages 210 ff). You may want to see how Fisher himself explained the difference (Fisher, 1958, pages 30 ff).

Now notice that from the definition of the mean, $\bar{Y} = p_1^2 Y_{11} + 2p_1p_2Y_{12} + p_2^2 Y_{22}$. Therefore using the genotypic values from Table 5-5, we have

$$p_1^2 a_{11} + 2p_1p_2a_{12} + p_2^2 a_{22} = 0. \qquad \text{5-8}$$

Since a_1 and a_2 are also measured as deviations from the population mean, we can follow the same procedure for the genic values:

$$p_1^2(2a_1) + 2p_1p_2(a_1 + a_2) + p_2^2(2a_2) = 0,$$

which is the same as

$$p_1a_1 + p_2a_2 = 0. \qquad \text{5-9}$$

TABLE 5-5

The genotypic model for the effect of alleles at a single locus on a multifactorial character.

GENOTYPE	A_1A_1	A_1A_2	A_2A_2
Frequency	p_1^2	$2p_1p_2$	p_2^2
Genotypic value	$Y_{11} = \bar{Y} + a_{11}$	$Y_{12} = \bar{Y} + a_{12}$	$Y_{22} = \bar{Y} + a_{22}$
Genic or additive value	$\bar{Y} + 2a_1$	$\bar{Y} + a_1 + a_2$	$\bar{Y} + 2a_2$
	$a_1 = p_1a_{11} + p_2a_{12}$	$a_2 = p_1a_{12} + p_2a_{22}$	

Genic and Genotypic Variances. I shall use lowercase-letter subscripts to indicate variances attributable to a single locus and capital-letter subscripts for variances of the character as a whole, to which many loci (and, in most cases, the environment) contribute. Since the values of a_{ij}'s represent deviations from the mean, we can write the genotypic variance directly from Table 5-5:

$$V_h = p_1^2 a_{11}^2 + 2p_1 p_2 a_{12}^2 + p_2^2 a_{22}^2. \qquad 5\text{-}10$$

We can similarly determine the genic variance from the quantities in Table 5-5:

$$V_g = p_1^2(2a_1)^2 + 2p_1 p_2(a_1 + a_2)^2 + p_2^2(2a_2)^2.$$

Now combining half of the first term and the first term in the expansion of the middle term, we obtain the first two terms in the expression below. Then follow the same procedure with the last term in the expansion and half the last term, which gives the third and fourth terms. The second line of the following equation represents what remains:

$$\begin{aligned} V_g = {} & 2p_1^2 a_1^2 + 2p_1 p_2 a_1^2 + 2p_1 p_2 a_2^2 + 2p_2^2 a_2^2 \\ & + 2p_1^2 a_1^2 + 4p_1 p_2 a_1 a_2 + 2p_2^2 a_2^2. \end{aligned}$$

The first two terms are $2p_1(p_1 + p_2)a_1^2$, but $p_1 + p_2 = 1$. We can similarly simplify the second two terms. Finally note that the three terms in the second line are twice the square of the left side of Equation 5-9 and therefore equal to 0. These simplifications lead to

$$V_g = 2(p_1 a_1^2 + p_2 a_2^2). \qquad 5\text{-}11$$

Look back at Equation 4-16. Note that Equation 5-11 is analogous to the numerator, which I called the variance of the excesses or the additive genetic variance. Equation 5-11 has the same interpretation, except here we are talking about the contribution of a locus to some quantitative measurement rather than to fitness.

Finally if we subtract V_g from V_h and call it V_d, we find (after a horrendous algebraic mess) that

$$V_d = [p_1 p_2(a_{11} - 2a_{12} + a_{22})]^2. \qquad 5\text{-}12$$

We use V_d to designate this expression because d symbolizes dominance. Notice that the term in parentheses is a natural measure of dominance: it measures the extent to which the heterozygote departs from the average of the two homozygotes. Thus we call V_d the dominance variance.

What we have done is to divide the total genotypic variance into two components. The first, the genic variance, is the variance attributable to additive effects of the two alleles. The second, the dominance variance, is a measure of the variance attributable to nonadditivity, or dominance.

We can extend this type of analysis with no essential change to multiple alleles. We can also extend it to two or more loci. In this case nonadditivity between loci creates effects analogous to dominance within a locus. The variance attributable to interlocus nonadditivity is called the **epistatic variance.** The extension of this analysis to epistatic interactions is given in C&K, pages 116 ff.

Covariance Between Relatives. Just as we obtained genic and genotypic variances, we can also write formulae for covariances between specified relatives. The information is set forth in Table 5-6.

The first two columns indicate all the possible genotypic combinations for two related individuals. The next two give the phenotypic deviations for each of these genotypes, from Table 5-5. The right section of the table gives the probabilities of these genotypic combinations if X and Y are (1) parent and offspring, (2) half-sibs, and (3) full sibs.

Consider parent and offspring, which always share one of the two alleles at the

TABLE 5-6

Calculation of the genetic correlation between parent and offspring, half-sibs, and full sibs.

GENOTYPE		PHENOTYPIC DEVIATION		FREQUENCY OF THIS COMBINATION, f		
X	Y	x	y	PARENT-OFFSPRING	HALF-SIBS	FULL SIBS
A_1A_1	A_1A_1	a_{11}	a_{11}	p_1^3	$\frac{1}{2}p_1^4 + \frac{1}{2}p_1^3$	$\frac{1}{4}p_1^4 + \frac{1}{2}p_1^3 + \frac{1}{4}p_1^2$
A_1A_1 A_1A_2	A_1A_2 A_1A_1	a_{11} a_{12}	a_{12} a_{11}	$2p_1^2p_2$	$2p_1^3p_2 + p_1^2p_2$	$p_1^3p_2 + p_1^2p_2$
A_1A_2	A_1A_2	a_{12}	a_{12}	$p_1p_2^2 + p_1^2p_2$	$2p_1^2p_2^2 + \frac{1}{2}p_1p_2$*	$p_1^2p_2^2 + \frac{1}{2}p_1p_2$* $+ \frac{1}{2}p_1p_2$
A_1A_1 A_2A_2	A_2A_2 A_1A_1	a_{11} a_{22}	a_{22} a_{11}	0	$p_1^2p_2^2$	$\frac{1}{2}p_1^2p_2^2$
A_1A_2 A_2A_2	A_2A_2 A_1A_2	a_{12} a_{22}	a_{22} a_{12}	$2p_1p_2^2$	$2p_1p_2^3 + p_1p_2^2$	$p_1p_2^3 + p_1p_2^2$
A_2A_2	A_2A_2	a_{22}	a_{22}	p_2^3	$\frac{1}{2}p_2^4 + \frac{1}{2}p_2^3$	$\frac{1}{4}p_2^4 + \frac{1}{2}p_2^3 + \frac{1}{4}p_2^2$

* Note that $p_1p_2 = p_1p_2^2 + p_1^2p_2$.

locus. In the first row one allele in X and one in Y are shared. The probability that the shared allele is A_1 is p_1. The other two alleles are independent (we are assuming Hardy-Weinberg ratios in the population); so the probability that they are both A_1 is p_1^2. The total probability is p_1^3. In the second row the shared allele is A_1. The probability that the shared allele is A_1 and that the unshared alleles are A_1 and A_2 is $p_1^2 p_2$. Since the combined genotypes of the second and third rows are equal, I have pooled their probabilities. The rest of the table is obtained the same way.

Putting all this information together with the definition of the covariance (the weighted average of the product of the deviations), we obtain

$$
\begin{aligned}
\mathrm{Cov}_{xy} &= p_1^3 a_{11}^2 + 2p_1^2 p_2 a_{11} a_{12} + p_1 p_2^2 a_{12}^2 + p_1^2 p_2 a_{12}^2 + 2p_1 p_2^2 a_{12} a_{22} + p_2^3 a_{22}^2 \\
&= p_1(p_1 a_{11} + p_2 a_{12})^2 + p_2(p_1 a_{12} + p_2 a_{22})^2 \\
&= p_1 a_1^2 + p_2 a_2^2 = \frac{1}{2} V_g .
\end{aligned}
\qquad \text{5-13}
$$

Thus the genetic covariance between parent and offspring is half the genic variance.

For half-sibs the method is essentially the same, and the probabilities are given in Table 5-6. Half-sibs share one allele half the time and none half the time. The first term in each expression is the probability when neither allele is shared (identical by descent); the second term is the probability when one allele is shared. Now multiply these expressions out, as we did for parent-offspring. One tip: sum each column separately and rewrite $p_1 p_2$ as $p_1 p_2(p_1 + p_2)$. Summing and simplifying (and remembering Equation 5-8), we obtain for the covariance of half-sibs

$$
\mathrm{Cov}_{HS} = \frac{1}{4} V_g . \qquad \text{5-14}
$$

You can probably guess the generalization of the covariance formula to other kinds of relatives. But first we need to note that the simple rule applies only to **unilineal** relatives, which share at most one allele that is identical by descent. Most relatives — half-sibs, parent and offspring, uncle and niece or aunt and nephew,* first cousins, one-and-a-half cousins, second cousins, and child and grandparent — are unilineal. The generalization is

$$
\mathrm{Cov}_{IJ} = 2F_{IJ} V_g , \qquad \text{5-15}
$$

where F_{IJ} is the coefficient of kinship of the two relatives.

---

* For these relationships I suggest the word *avuncles,* from the Latin *avunculus.* Curiously the adjective avuncular exists in English, but not the noun.

Full sibs are more complicated, since they are **bilineal:** they can share two different alleles that are identical by descent. When two alleles are shared, the covariance includes the dominance variance.

Look again at Table 5-6. Full sibs share no alleles with probability 1/4, one allele with probability 1/2, and two alleles with probability 1/4. The three terms in each expression correspond to these events. Again consider the terms separately. The first terms, when multiplied by the phenotypic deviations x and y, sum to 0. The second terms sum to $V_g/4$. The third terms sum to $V_h/4$. Putting these sums together, we have $V_g/2 + V_h/4$, which, since $V_h = V_g + V_d$, gives

$$\mathrm{Cov}_{FS} = \frac{1}{2} V_g + \frac{1}{4} V_d \qquad \text{5-16}$$

for the covariance between full sibs.

Extending Results to Multiple Loci. As long as the individual loci that contribute to the trait are in linkage equilibrium and are additive in their effects, the different types of variances and covariances are additive. Letting a capital letter stand for the sum of the variances or covariances at all relevant loci, we have

$$\mathrm{Cov}_{IJ} = 2F_{IJ}V_G \qquad \text{5-17}$$

and

$$\mathrm{Cov}_{FS} = \frac{1}{2} V_G + \frac{1}{4} V_D. \qquad \text{5-18}$$

Most of the time we use such covariances to predict the result of artificial selection. The breeder uses observed covariances between unilineal relatives to measure the genetic covariances. Half-sibs with the same father but different mothers are usually used, since the breeder can easily randomize the environments and minimize maternal effects. Since for half-sibs $F_{IJ} = 1/8$, from Equation 5-17 we have that V_G is four times the measured covariance. Then from this value the breeder would like to predict the result of selection.

For this prediction we need the expected measurement of the offspring, given the average measurement of the parents. In technical parlance we need the regression of the offspring on the mid-parent. Recall that the regression is the ratio of the covariance to the variance of the parental average.

First, the variance of a single parent is V_P, the phenotypic variance, which includes both genetic and environmental effects. The mean of n values is $1/n$ times the variance of any one value (see Table 5-3, row 7). Thus when $n = 2$, the variance of the parental average is 1/2 the phenotypic variance of a single parent.

The covariance of the offspring and the average of the parents turns out, somewhat surprisingly, to be the same as the covariance of the offspring and a single parent. We can derive this relationship as follows, where P_f and P_m are individual female and male parents and O is an individual offspring; overbars indicate averages.

$$\begin{aligned} \mathrm{Cov}_{\bar{P}O} &= \frac{\Sigma(1/2)\,(P_m + P_f - \bar{P}_m - \bar{P}_f)\,(O - \bar{O})}{N} \\ &= \frac{1}{2}\,\frac{\Sigma(P_m - \bar{P}_m)\,(O - \bar{O})}{N} + \frac{1}{2}\,\frac{\Sigma(P_f - \bar{P}_f)\,(O - \bar{O})}{N} \\ &= \mathrm{Cov}_{PO}. \end{aligned} \tag{5-19}$$

From this equation we can write the regression of offspring on mid-parent as

$$\begin{aligned} b_{O\bar{P}} &= \frac{\mathrm{Cov}_{O\bar{P}}}{V_{\bar{P}}} \\ &= \frac{2\mathrm{Cov}_{OP}}{V_P} \qquad \text{(from Equation 5-19 and Table 5-2)} \\ &= \frac{V_G}{V_P} \qquad \text{(from Equation 5-13)} \\ &= H_N, \end{aligned} \tag{5-20}$$

where H_N is the **narrow-sense heritability.** We thus arrive at the prediction equation given earlier in words (see Equation 5-1):

$$O = M + H_N(\bar{P} - M). \tag{5-21}$$

This theory is based on a single-locus model. However, if the alleles at different loci are additive in their effects and in linkage equilibrium, we can simply add up the effects of the single loci. Thus the single-locus model leads to a general theory of quantitative inheritance.

Let us review our assumptions. First and most important, we assumed that the genetic and environmental influences are independent and additive (or that the variables have been transformed so that they are additive). Genotype-environment interactions are often important, and we shall return to this topic later in the chapter.

Second we have ignored epistasis. Most of the time epistasis does not make a large contribution to correlations and regressions of unilineal relatives. The reason is that, as they enter the formulae, they are accompanied by small coefficients (1/4 or less for parent and progeny, 1/16 or less for half-sibs), which diminish their influence. In practice, epistasis is usually unmeasurable by animal and plant breeders and is often ignored in selection experiments.

Linkage equilibrium is also a safe assumption for multifactorial traits. Most organisms of commercial interest have a large number of chromosomes, so that any pair of loci contributing to a trait are usually unlinked; if they are on the same chromosome, they are not likely to be close enough for appreciable linkage equilibrium. Clusters of linked alleles, such as HLA or beta globins, are best treated as a single locus. Therefore in conventional breeding practice, linkage is usually ignored.

5-6 HERITABILITY IN THE NARROW SENSE

We have seen how to estimate narrow-sense heritability and use it to predict the results of selection. We can extend Equation 5-4 by separating the genotypic deviation into additive and dominance parts:

$$p_i = g_i + d_i + e_i, \qquad 5\text{-}22$$

where

g_i = the genic, or additive, deviation

d_i = the dominance deviation.

These deviations are diagrammed in Figure 5-9.

G_i is often called the **breeding value.** Since g_i/p_i is the narrow-sense heritability, H_N, the breeding value of an individual is

$$G_i = M + H_N(P_i - M). \qquad 5\text{-}23$$

FIGURE 5-9. A further subdivision of genetic and environmental deviations.

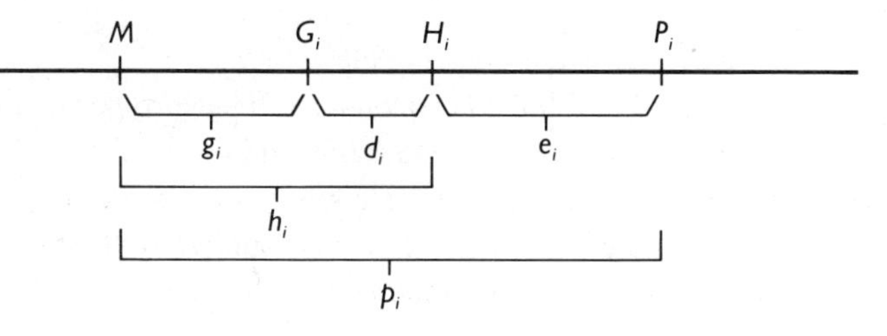

Therefore we can also state the prediction formula as follows:

> **The predicted phenotype of the progeny is the average of the breeding values of the parents.**

Estimating the Narrow-Sense Heritability. The usual way to estimate H_N is to observe the correlation between unilineal relatives. If we divide both sides of Equation 5-17 by the phenotypic variance, we have

$$r_{IJ} = 2F_{IJ}H_N$$

or

$$H_N = \frac{r_{IJ}}{2F_{IJ}}, \qquad \text{5-24}$$

where r_{IJ} is the measured correlation between relatives I and J. Therefore the rule for estimating the narrow-sense heritability is to divide the measured correlation by twice the kinship coefficient.

In dairy cattle the breeding value of a bull is often measured directly. The breeder mates the bull (usually by artificial insemination) to a large number of randomly chosen cows. Since the expected performance of the daughters is the average of the breeding values of the parents, and the population average is M, the breeding value of the bull is twice the amount that his daughters deviate from the population average (or the average of their mothers). With artificial insemination one male can have thousands of progeny; so there is great importance in assessing the breeding value accurately and using those bulls with the highest value.

Continuation of the Guinea Pig Example. Looking again at the data in Table 5-4, we can now include some more information. Wright also estimated the correlation of the amount of white spotting between relatives. Within inbred lines this correlation was practically 0, as it should be. In the randomly mating population the estimate of narrow-sense heritability, based on correlation between relatives, was 0.380 (see Table 5-7). Thus V_g is (0.380)(573), or 218. We can fill in the rest of the table by subtraction. Because this trait shows very little dominance, the broad- and narrow-sense heritabilities are essentially the same.

We cannot use full-sib correlations to measure heritability for two reasons. The first and most important, as we saw in Section 5-5, is that full-sib covariances include part of the dominance variance. Thus full-sib correlations overestimate the additive variance and hence the narrow-sense heritability. At the same time they do not include all the dominance variance, which makes them unsuitable for measuring broad-sense heritability.

TABLE 5-7

A further analysis of the data in Table 5-3.

COMPONENT	VARIANCE	FRACTION OF TOTAL
V_G	218	0.380
V_D	15	0.027
V_H	233	0.407
V_E	340	0.593
V_P	573	1.000

The second reason is that full sibs, especially mammals, almost always have environmental correlations caused by having the same mother. The correlations may be prenatal or early postnatal, since the young cannot be separated from the mother too early. For both of these reasons the breeder prefers half-sibs with the same father but different mothers.

Correlations for both physical and behavioral traits between human sibs often exceed parent-child correlations. Because the effects of dominance and environmental similarity are confounded, we usually cannot determine the cause of this correlation increase.

What About Epistatic Variance? We can apply the same principles used to separate dominance from the additive variance to take epistasis into account. A small amount of epistatic variance is included in the correlations of even unilineal relatives, and in practice we usually cannot estimate it. Fortunately, as mentioned earlier, most of the evidence suggests that epistasis does not greatly influence most quantitative characters of agricultural plants and livestock, and ignoring it in breeding experiments does not lead to serious errors. A more serious complication is interaction, not between genes but between genotype and environment. This interaction is the subject of the next section.

5-7 INTERACTION AND COVARIANCE

Finally we come to some of the troublesome realities in measuring and applying heritabilities, especially where human data are concerned. The first problem is that variances, covariances, and heritabilities are subject to large statistical errors. Large amounts of data are required to generate reproducible values. This problem is not too serious in plant breeding because large numbers are usually available. Such numbers are not so readily available with farm animals, but usually breeders pool data from several stations. Human data that are suitable for heritability analysis are often quite limited.

The second problem comes from genotype-environment interaction and covariance. A more detailed version of Equation 5-4 is

$$p_i = h_i + e_i + i_i, \qquad 5\text{-}25$$

where i stands for interaction between h and e. Interaction is nonadditivity of h and e.

In terms of variance components

$$V_P = V_H + V_E + V_I + 2\mathrm{Cov}_{HE}. \qquad 5\text{-}26$$

The last two terms are the variance due to interaction and twice the covariance of heredity and environment. If you are puzzled as to the origin of this last term, see Equation A-8 in the appendix. Let us now consider the sometimes subtle and confusing distinction between interaction and covariance.

Interaction. Interaction occurs when the effects of genotype and environment are not additive; that is, when the environment has a different effect on different genotypes. Some strains of corn grow better in sandy soil, others better in clay; sugar is harmful to a child with hereditary diabetes, but not to a normal child; a musically talented child profits more from a conservatory education than a child that is tone-deaf.

In livestock and plant breeding, interactions are often small: an individual that does well in one environment does well in others that it is *likely* to encounter (but not in a greatly different environment; a high-producing dairy cow would not survive long in the jungle). Sometimes the interaction is important and can be incorporated into the breeding strategy. In the development of corn hybrids, there are genotypes that do poorly in all environments; they are quickly discarded. But among the best hybrids some may be better suited to a long growing season, some to a dry climate, some to a sandy soil, and so on. By choosing particular strains for particular environments, the breeder takes advantage of genotype-environment interactions. In a sense (if this is not using pompous words to describe a simple idea), much of medical care is the judicious use of genotype-environment interactions: a drug that is useful in one genotype may be useless or harmful (for example, allergic) in another. However, the environmental causes of most quantitative traits are not sufficiently understood to make interaction more than a statistical abstraction or an impediment to a simple interpretation.

Sometimes we can remove or reduce interaction by a suitable transformation. For example, if the genotype and environment are multiplicative rather than additive, a logarithmic transformation will remove the interaction. Often if a log transformation removes or reduces epistasis, it does the same for genotype-environment interactions. However, much of the time the situation is too complicated to be solved by a simple transformation of data.

Covariance. Genotype-environment covariance arises when genotypes are not randomly allocated to different environments. A farmer creates a covariance by feeding the best cows better than the worst. Sending musically talented children to a conservatory and the tone-deaf to a school that emphasizes nonmusical subjects creates a covariance. Although this strategy is undoubtedly beneficial to the child, it is not a good design for an experiment to separate genetic and environmental variances.

In planned experiments covariance is not a problem; it is removed by proper randomization. But in human studies such randomization is not possible, and covariances usually complicate the interpretation.

Heritability of Human Traits. Until now we have considered heritability theory and its applicability to physical traits and to animal and plant breeding. These applications are not controversial for the most part, although opinions do differ as to the best interpretation of covariances and interactions and in the minutiae of calculation procedures. Human traits, especially behavioral traits, are usually hard to study, and the interpretations of such studies are often emotion laden as well. This is particularly true in studies of the heritability of intelligence, as measured by IQ.

Correlations between human relatives usually lack a clear intepretation, because relatives tend to live in similar environments. Thus genetic correlations are confounded with environmental correlations. Only special circumstances, such as the comparison of unrelated children reared by the same adopting parents or related children reared in different homes, provide the requisite separation. Even so, there is doubt that the data provide in practice what is expected in theory.

In theory we should easily be able to interpret the study of separated identical twins. Their correlation should provide a direct measure of broad-sense heritability. One problem, however, is that the numbers are small. Altogether the four largest published studies of identical twins include 122 pairs: in some cases the twins were not separated early; in other cases they were reared by persons who were related or were near neighbors; and in some cases one twin was reared by its own parents. Also, do adopting parents constitute a random sample of the population? Are twins who are separated in infancy a random sample of twins? Do twins, because of their common intrauterine environment (for example, crowding), differ in subtle, but important, ways from other children? Finally the largest and, in principle, the best study, that of Cyril Burt in London, has so many internal inconsistencies and even evidence of fraud that the data cannot be accepted.

In a classical study back in the 1930s, H. H. Newman studied 19 pairs of one-egg twins that had been separated at an early age and reared in separate homes. At face value the measured heritability was about 0.7 (see Newman, Freeman, and Holzinger, 1937). Correcting for errors of measurement would raise the heritability value to about 0.75. On the other hand, environmental correlations would lower the value. The data from other twin studies and studies of adopted children (reviewed in Wright,

1978, volume IV) suggest that the value is probably higher than 0.6, but a more conservative view is that the data are so scanty and depart so far from the requirements of a good scientific experiment that any numerical estimate of the heritability of IQ is unwarranted.

The Scandinavian countries provide additional information on IQ heritability. In Denmark, for example, draft board examinations include IQ tests. This information and an extensive record of adoptions led to the opportunity to study half-sibs and full sibs reared in different homes and unrelated children reared in the same home. The data suggest a high narrow-sense heritability with very little dominance, but the numbers are too small for definite conclusions at this stage. Home and educational environments in Denmark, especially homes with adopted children, may be quite similar. Therefore we should perhaps expect a higher heritability in Denmark than in the larger, more diverse United States, where environments are more variable.

5-8 ASSORTATIVE MATING

In Chapter 2 we saw that assortative mating causes a change of heterozygosity, albeit a small one. It also causes an increase in variance, although I did not emphasize it at the time. Assortative mating is qualitatively like inbreeding, but its effect on variance is greater than its effect on homozygosity, as I shall show later in the chapter.

We need to distinguish between two kinds of inbreeding effects. If the population is large but has some consanguineous matings, the homozygosity will increase, as will the variance. Yet the cumulative effect is very small, unless those individuals who are the result of consanguineous matings preferentially mate with their relatives. On the other hand, if the pattern of inbreeding is such that the population breaks up into groups, the situation is quite different. In this case the homozygosity increases within the group, and the variance for additive genes within the group decreases (it is proportional to $1 - F$). The whole population becomes more homozygous, but different alleles will tend to be fixed in different subpopulations. The whole population variance increases because of the divergence of the groups (for additive genes the population variance increases in proportion to $1 + F$).

Assortative mating based on phenotype does not usually cause the population to break up into groups. A population in which each individual chooses a mate similar to itself (of the same size, for example) will become more variable but will not break up into discrete size groups.

Various Causes of Assortative Mating. One cause of assortative mating is that it is an indirect consequence of inbreeding. If mating individuals are related, they will also be correlated in size and other genetically determined measurements. Part of the assortative mating in the human population is undoubtedly caused by common ancestry: individuals from the same group tend to mate within the group.

A second reason is phenotype choice. Large individuals may choose large mates, regardless of the cause of the size. In the human population height is highly correlated between husband and wife, as is weight.

A third reason is that the trait may be associated with another trait for which mating is assortative. For example, there is a sizable correlation for length of forearm between husband and wife. A mate choice based on this trait is hard to imagine. The choice is much more likely to be based on height or overall size. Any assortative mating for height will be reflected in a similar assortment for forearm length, diminished somewhat by the lack of perfect correlation between the two traits. In British populations about the turn of the century, the marital correlation for height was 0.28 and for forearm length 0.20. The lower correlation for the latter implies that it is of secondary importance in the choice of a mate, which is what we would suspect.

The specific cause of the assortative mating influences the genetic analysis. The more closely the assortment is based on the genotype, the greater the genetic consequence. Thus if the assortative mating is caused by common ancestry, the consequences are greater than if it is based on purely phenotypic choice, unless the heritability is complete. If the trait were purely environmental, assortative mating of course would have no genetic consequence.

From here on we shall consider assortative mating to be based on phenotype only.

Comparison of Inbreeding with Assortative Mating. As I said, both inbreeding and phenotypic assortative mating increase homozygosity and variance, but quantitatively their effects are different.

> **Inbreeding has a greater effect than assortative mating on homozygosity, whereas assortative mating for a quantitative trait causes a greater increase in the variance.**

A simple example will clarify the variance-enhancing effect of assortative mating. Suppose that two loci, both without dominance, cumulatively affect a trait. Suppose each allele with subscript 1 adds 1 unit to the character (for example, size). Thus $A_1A_1\ B_1B_1$ and $A_0A_0\ B_0B_0$ represent extreme phenotypes. Table 5-8 illustrates the model.

Inbreeding increases all homozygous types; assortative mating increases only the phenotypic extremes. This simple example in Table 5-8 shows that the variance is doubled at equilibrium with complete homozygosity. This is expected, since the total variance increases as $1 + F$, and with complete homozygosity $F = 1$. The assortative mating increase is fourfold, twice as great as with inbreeding. This increase is for the two-locus case; if n loci were involved, the increase over the random mating value would be $2n$-fold. This example gives the qualitative picture: assortative mating can cause a large increase in the variance of traits determined by many loci.

Because assortative mating is never perfect, the population never separates into two extreme groups, as Table 5-8 might seem to imply. It just becomes more variable. The general equilibrium formulae, although difficult to derive, are easy to understand (for derivations see C&K, pages 99 ff and pages 148 ff). The formulae are

$$\text{Inbreeding} \qquad \hat{V} = V_0(1 + \hat{F}) \qquad \text{5-27}$$

$$\text{Assortative mating} \qquad \hat{V} = \frac{V_0}{(1 - r(1 - 1/2n)}, \qquad \text{5-28}$$

where V_0 is the variance before inbreeding or assortative mating started.

In these formulae $\hat{F}$ is the inbreeding coefficient at equilibrium, n is the number of loci, and r is the correlation between mates; r is assumed to be constant during the time required to reach equilibrium. Note that these formulae hold for additive genes only. However, they are good approximations, since alleles contributing to polygenic traits are usually nearly additive.

Notice that, although there may be a large increase in variance, there is not much change in homozygosity. This lack of change in homozygosity is because, although assortative mating puts together genes with similar effects on the phenotype, these genes are usually not alleles.

Assortative mating and inbreeding differ in another way. We have seen before that

TABLE 5-8

A simple two-locus model illustrating the difference between inbreeding and assortative mating. Alleles with subscript 1 have frequency p; those with subscript 0 have frequency q.

		EQUILIBRIUM FREQUENCY		
GENOTYPE	CODED PHENOTYPE	RANDOM MATING	COMPLETE INBREEDING	COMPLETE ASSORTATIVE MATING
$A_1A_1\ B_1B_1$	$2y$	p^4	p^2	p
$A_1A_1\ B_1B_0$, $A_1A_0\ B_1B_1$	y	$4p^3q$	0	0
$A_1A_1\ B_0B_0$, $A_1A_0\ B_1B_0$, $A_0A_0\ B_1B_1$	0	$6p^2q^2$	$2pq$	0
$A_1A_0\ B_0B_0$, $A_0A_0\ B_1B_0$	$-y$	$4pq^3$	0	0
$A_0A_0\ B_0B_0$	$-2y$	q^4	q^2	q
Variance		$4pqy^2$	$8pqy^2$	$16pqy^2$

one generation of random mating erases all previous effects of inbreeding. This effect does not occur with assortative mating, which creates gametic disequilibrium by putting together alleles with similar effects in the same gamete. Hence the increased variability associated with assortative mating disappears with random mating only at the rate that linkage equilibrium is approached.

Effects of Dominance and Environment. The formulae of the previous section are appropriate for a trait with heritability 1. What happens with complications of dominance, epistasis, and environment?

The effect of assortative mating for multifactorial traits is to increase only (or mainly) the additive component of the variance. Dominance variance is hardly changed at all, and epistatic variance (which is usually small anyhow) changes very little. Likewise unless the environment has changed or there is genotype-environmental covariance or interaction, the environmental variance stays the same.

We would expect the additive variance to increase, as given in Equation 5-28, except that we should replace r by the genic (or additive genetic) component of the correlation, which I shall call A. A is rH_N. Thus to incorporate dominance and environment, we assume n is large and replace Equation 5-28 by

$$\hat{V}_P = \hat{V}_G + V_D + V_E$$
$$\hat{V}_G = \frac{V_G}{1 - \hat{A}}, \qquad \hat{A} = \hat{H}_N r. \tag{5-29}$$

A circumflex over the letter indicates the equilibrium value. The other letters represent the values before assortative mating began.

Considerable assortative mating takes place in the human population, especially for educational attainment and IQ. To whatever extent IQ is genetic, this should increase the population variability. Furthermore environmental correlations work somewhat similarly. It is hard to estimate how great these variance-enhancing effects are, but they may be substantial.

Assortative Mating and Animal Breeding. Since the rate of progress of selection depends on the genic variance, we would expect that assortative mating, by increasing the variance, would increase the rate of improvement of plants and animals by selection. Our expectation is correct.

Although assortative mating does not alter the prediction Equation 5-21, it increases the effectiveness of selection in two ways:

1. **Since assortative mating increases the genic variance without changing the other variance components much, it increases the heritability.**

2. **Since the population is more variable, the mean of the selected parents is increased; that is, the "reach" of the breeder is increased. Coupling assortative mating with selection can enhance the rate of progress.**

5-9 STABILIZING SELECTION

In nature the fittest phenotype is almost always quite close to the population mean. A mouse that is too small or too large is less viable and fertile than one that is near the average. Nature seems to abhor outliers almost as much as it does a vacuum.

In addition to everyday observations, several studies have documented the relationship of fitness to proximity to the mean. One classical study showed that the farther a newborn human infant is from the average weight, in either direction, the less is its chance of survival. Another investigator took advantage of a severe storm to measure the size of the sparrows that had been killed. These sparrows turned out to be more variable than the normal population.

We have just seen that directional selection can very rapidly change the size of mice. Yet mice have been the same size for tens, or hundreds, of millions of years. Even a tiny consistent directional selection for increased size would have long since transformed them to the size of an elephant.

The conclusion is inescapable:

A great deal of natural selection is directed against those individuals that deviate from the mean.

This conclusion leads to a paradox. Look again at the model of Table 5-8. Suppose that selection continues in favor of coded phenotype 0. Intuition tells us that this phenotype will increase, but will the population tend to become a homozygous type, such as $A_1A_1\ B_0B_0$, or the heterozygote $A_1A_0\ B_1B_0$?

We have the following hint: A population that is made up entirely of one of the two favorable homozygous types is stable, and all individuals are of maximum fitness. On the other hand, heterozygotes in the population will always produce unfavored types through Mendelian recombination.

We can look at the question another way: Gametes $A_1\ B_0$ and $A_0\ B_1$ are favored because they will more often produce zygotes of intermediate phenotype, which is favored by natural selection. Suppose that for some reason, random drift perhaps, $A_1\ B_0$ is more common. Either $A_1\ B_0$ or $A_0\ B_1$ will combine with the common type to produce an optimum phenotype, but the $A_0\ B_1$ chromosome will produce gametes by recombination that lead to less-favored genotypes in the next generation. Thus we expect that the A_1 and B_0 alleles would increase and that the population would eventually become homozygous.

Extending this reasoning to a larger number of loci, we conclude that natural selection tends to make the population homozygous for some genotype that is at the optimum. The theory was first worked out by Fisher, Haldane, and Wright, as usual. If there is partial or complete dominance, the situation is more complicated. One article you might consult is Kojima (1959). The mathematics is difficult, but the conclusion is entirely reasonable: selection tends toward homozygosity. However, if the optimum phenotype lies between the two nearest homozygotes, one locus may be left segregating, and its allele frequencies will make the population mean coincide with the optimum. The larger the number of loci involved in the trait, the smaller the fraction of loci left segregating.

The paradox that I referred to is that quantitative traits nearly always exhibit considerable variability, despite selection pushing toward homozygosity and therefore uniformity. What causes the variability?

One possibility is that genes for quantitative traits have pleiotropic effects. The universality of pleiotropy is well established. However, we must assume that some of the pleiotropic effects show overdominance, leading to the kind of stable polymorphism discussed in Section 4-7. The problem is that no evidence supports such ubiquitous overdominance.

A more conventional explanation is that the variability is simply the result of mutation. Figure 5-10 shows a popular model. We can reasonably assume that

FIGURE 5-10. A diagram of the quadratic optimum model. The mean phenotype is at, or very close to, the optimum. The fitness decreases in proportion to the square of the distance from the optimum. M is the mean phenotype, d is the deviation from the mean, and O is the maximum fitness.

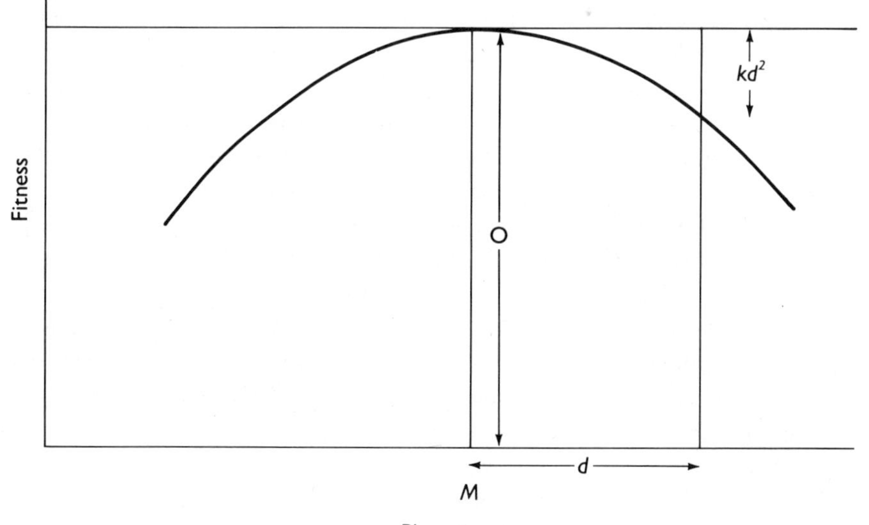

phenotypes close to the mean have the highest fitness and that the decline in fitness accelerates as the phenotype departs from the average. A range of values near the optimum are often of nearly the same fitness, but that fitness decreases rapidly once the departure from the optimum becomes large. Therefore we expect that the fitness distribution would be humped, as the figure shows. The simplest realistic model is the **quadratic optimum model,** in which we assume that the fitness declines in proportion to the square of the deviation, d, from the optimum phenotype. In symbols, we have

$$\text{Fitness} = \text{optimum fitness} - kd^2. \qquad \text{5-30}$$

We can think of mutations as either increasing or decreasing the value of the phenotype (for example, size). A mutation that increases size will be favored in a small individual but will be deleterious in a large one. Thus the average effect of a mutant will be very small, since it is sometimes beneficial and sometimes deleterious. Since mutation is opposed by such weak selection, the mutation-selection equilibrium will be a high frequency of the mutant allele. Hence a low input of mutations is sufficient to maintain a great deal of variability on the phenotypic scale.

Extending Equation 5-30, we note that the difference between the optimum fitness and the mean fitness of the population will be

$$L = \sum P_i k d_i^2 = k \sum P_i d_i^2, \qquad \text{5-31}$$

where P_i is the frequency of individuals of phenotype d_i. I have used L for this expression because it is analogous to the load used in Section 4-5, which is the amount that mean fitness is reduced by variability for this phenotype.

Experience tells us that measured traits tend to be, or can be transformed to be, approximately normally distributed. Furthermore the mean of the population is quite close to the optimum. If the mean and optimum coincide, then L is simply k times the variance of the phenotype. The effect of mutation is to increase the variance and hence the load. The deleterious consequences of mutation are like those discussed in Chapter 4. In this case, however, the deleterious effect is to produce more individuals that deviate from the optimum. As mentioned before, since the effect of an allele depends on the size of the individual in which it occurs, the average deleterious effect of a single mutant gene is very small.

Pleiotropy and epistasis can also lead to a stable equilibrium. As a simple model consider two loci without dominance, each affecting the same two traits. At the first locus one allele affects both traits positively and the second affects both traits negatively. At the second locus one allele affects the first trait positively and the second negatively while the other allele has the reverse effect. This is not unreasonable, for it is quite likely that with limited resources an increase in one function leads to a decrease in another. In such a system, A. Gimelfarb (*Genetics,* 1986, in press) showed that with an intermediate optimum both alleles are maintained at both loci, hence increasing the population variability. This, and more complicated variations on the same theme, may well be an important factor maintaining population variability.

The overall picture that emerges from this discussion is that almost every quantitative trait in any species has an intermediate optimum. Natural selection keeps the population mean near this optimum. Mutation (and possibly overdominant epistatic pleiotropic effects) maintain variability. The phenotype contains a great deal of additive genetic variance. This feature allows the population mean to move quickly to a new optimum whenever the environment changes. The slow change of the horse from an animal the size of a fox to its present size is not the result of directional selection, as the animal breeder thinks of it, but of continual stabilizing selection toward a slowly changing optimum.

QUESTIONS AND PROBLEMS

1. The narrow-sense heritability of growth rate in swine is 0.4, the broad-sense heritability is 0.6, and the population mean size is 100 kg. What is the expected weight of the progeny of a 75-kg female and a 105-kg male?
2. The phenotypic variance of yield in maize is 200 bushels^2 per acre. The variance within an inbred line ($F = 1$) is 80. The correlation between half-sibs is 0.08. What are: (a) the genic variance? (b) the genotypic variance? (c) the environmental variance? (d) the narrow-sense heritability? (e) the broad-sense heritability?
3. Assume that the mean height of human males is 170 cm, the narrow-sense heritability is 0.6, and the broad-sense heritability is 0.8. (a) A man is 180 cm tall. How tall would we expect him to be if he had been reared in an average environment? (b) This man marries a woman of average height, and their son is adopted into an average home. What is the expected height of the son (when he grows up)? (c) What is the expected height of the son if he is reared in the same environment as his father?
4. Two homozygous strains of potatoes yield 10 and 100 bushels per acre. The F_1 yields 20 bushels per acre and the F_2 yields 25. Test these values for consistency with additive and multiplicative gene action.
5. Derive the variance formulae in Table 5-8.
6. Would you believe someone who told you that the correlation in intelligence between half-sibs is 0.3?
7. Under what circumstance might the phenotypic variance be less than the sum of the genotypic and environmental variances?

8. (a) Are double first cousins unilineal or bilineal? (b) Sketch a pedigree of bilineal relatives whose kinship coefficient is 1/16.

9. (a) What is the most likely explanation of a higher correlation between half-sibs with a common mother than between those with a common father? (b) What is the most likely explanation if the two kinds of half-sib correlations add up to less than the full-sib correlation?

10. The narrow heritability of human height in a randomly mating population is 0.7. The population now mates assortatively, with the correlation between parents being 0.28. After equilibrium the heritability is 0.79. How much has the genic variance increased?

11. What is the heritability of sex?

12. Would you expect the selection for oil content to have been as effective if corn were asexual?

13. The genic variance is 60, the environmental variance is 20, and the total variance is 100. What is the correlation between full sibs?

14. Which has the greater heritability, a trait determined by a rare dominant allele or one determined by a rare recessive allele?

15. What is the prediction formula for selection in an asexual population?

16. (a) Why doesn't the variance of a polygenic trait decrease with selection (or at least not decrease much at first)? (b) Would this situation occur in an asexual population?

17. Why are full sibs unsuitable for measuring either the broad- or the narrow-sense heritability?

18. Using the parameterization of Table 4-3, show that the narrow-sense heritability approaches 1 as h approaches 0 or 1 and is minimal when $h = q$.

CHAPTER 6

POPULATIONS WITH OVERLAPPING GENERATIONS

In the earlier chapters we regarded populations as having discrete, nonoverlapping generations. This model is precisely correct only in restricted cases, such as annual plants and 17-year cicadas. Yet as we saw from laboratory and field observations, the model provides a satisfactory approximation for more complex situations. The human population is in Hardy-Weinberg proportions for many loci that have been studied; most departures are accounted for by differential mortality before the age of enumeration. Formulae based on the simplified model work very well for determining the effects of inbreeding and random drift and for predicting the results of selection in breeding experiments. Such formulae are especially suitable if our interest is in equilibria, where it makes little difference whether the generations are distinct or not. Therefore for most purposes the simplified models of the previous chapters work as well as the accuracy of the data deserve and, as we have seen, they have the enormous advantage of leading to simple, easily understood mathematical formulations.

On the other hand, if we want a more detailed look into population changes and evolution in organisms in which death and reproduction occur at various ages, we need to take the life history into account. This is particularly important if we want to consider the evolution of the life cycle itself.

When we try to develop formulae for an age-structured population that are analogous to those in earlier chapters for allele frequency change and inbreeding, we immediately encounter almost insurmountable mathematical complexities, which are certainly beyond the scope of this book. For this reason I shall ignore Mendelian inheritance for now and treat populations as if the complications of meiosis, fertilization, segregation, and linkage were nonexistent. We regard the population as if it were either asexual or genetically uniform. Despite this regrettable necessity, we shall discover many interesting principles. They apply to ecological problems, such as the relationships between different species (as opposed to the geneticist's preoccupation with differences within a species), and we can carry qualitative conclusions over to genetic situations, as I shall discuss later in the chapter.

Our first job is to formulate a model.

6-1 AGE-STRUCTURED POPULATIONS OBSERVED AT DISCRETE INTERVALS

As a simple model of a population with age structure, assume that the organism has a specific breeding season. To be concrete, assume that all births occur on the same day, say July 1, and that individuals may die at any time but may also survive to reproduce for several years. In this model each individual necessarily reproduces only on its birthday anniversary at 1, 2, 3, . . . , years of age. We should census the population immediately after the births occur before any newborns have died. Figure 6-1 illustrates this population.

This model differs from the Leslie model, which is often given in ecology textbooks. In the Leslie model the census is made just before the time of birth, and only those newborns that have survived the first year are counted. This model is appropriate for many animal species in which newborns are difficult to observe and the interest lies in those that survive until the next enumeration period. The method I have used is appropriate in cases in which newborns are counted and the interest is in the whole population, such as in the human population. The differences are matters of detail and become decreasingly important as the interval between censuses is decreased.

In species with two sexes we need to adopt some convention as to how to allocate progeny to the two parents. If we know the parentage, we can credit half of each offspring to each of its parents; but we do not know the paternity in many animals. Even if we do know it, do we count the male's contribution according to his age at the time of conception or at the time of birth of the progeny? Although such questions are good for

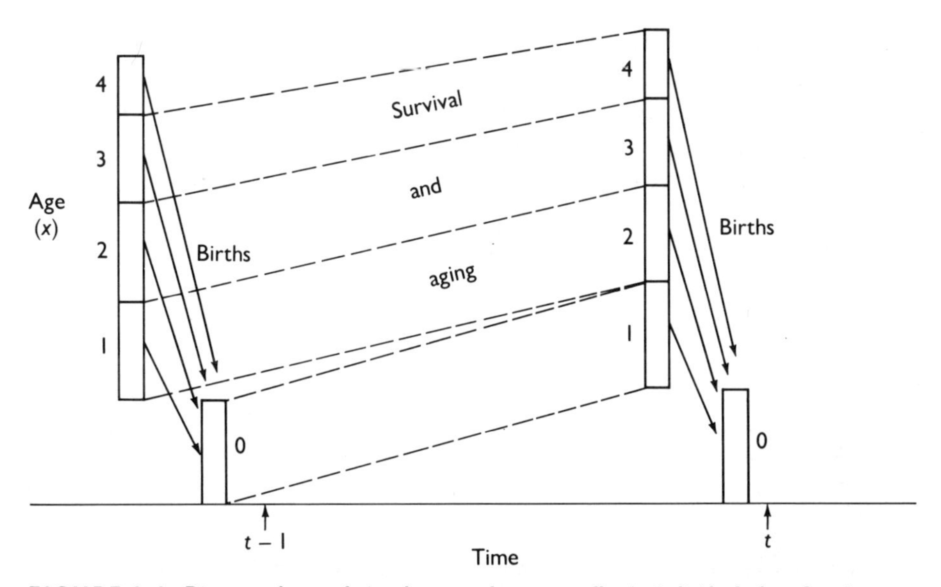

FIGURE 6-1. Diagram of a population that reproduces annually. An individual of age 1 at time t was age 0 at time $t - 1$, an individual of age 2 was age 1, and so on. Reproduction takes place on the birthday anniversary of each individual, and the population is enumerated immediately after. At the next census the number of each group aged 1 year and over will depend on the number aged 1 year less at the previous census multiplied by the probability of survival through the year. The number of newborns, age 0, is determined by the numbers of each group and the age-specific fertilities, b_x.

demographers to grapple with, we shall simply avoid the issue, as is customary, by counting only females in each generation. Of course, if we know the sex ratio at each age, we can easily determine the total population number from data on females only. Alternatively, think of the population as being asexual.

Let n_{xt} be the number of females of age x at time t, both measured in years and counted immediately after the breeding season. Let p_x be the probability of surviving from age x to age $x + 1$. Let b_x be the number of births to a female at age x. Finally, let $q_x (= 1 - p_x)$ be the probability of death in the age interval x to $x + 1$. We shall regard the quantities p_x, q_x, and b_x as constants.

The number of females of age 10 in the year 1980 will be the number who were age 9 in 1979 and have survived for the ensuing year. Thus if m is the maximum age of survival,

$$n_{xt} = n_{x-1,t-1} p_{x-1} \qquad \text{where } x = 1, 2, 3, \ldots, m, \qquad n_{m+1} = 0. \tag{6-1}$$

Likewise the number of newborns at time t will be

$$n_{0t} = n_{1t} b_1 + n_{2t} b_2 + \ldots + n_{kt} b_t = \sum_{x=1}^{k} n_{xt} b_x, \tag{6-2}$$

where k is the maximum age of reproduction ($k \leq m$). With these equations we can calculate the population number at each age for the next year and for successive years.

We can perhaps clarify the relationships by considering a hypothetical species in which no individual survives to the sixth year or reproduces after the fourth ($m = 5$, $k = 4$). If we start at time $t = 0$, the numbers at time $t = 1$ are given by

$$
\begin{aligned}
n_{01} &= n_{11}b_1 + n_{21}b_2 + n_{31}b_3 + n_{41}b_4 \\
n_{11} &= n_{00}p_0 \\
n_{21} &= n_{10}p_1 \\
n_{31} &= n_{20}p_2 \\
n_{41} &= n_{30}p_3 \\
n_{51} &= n_{40}p_4 \\
n_{61} &= 0.
\end{aligned}
$$

Table 6-1 gives a numerical example. Each line shows the number in each age group and the total population number. The population starts out, somewhat artificially, with 10,000 newborns at $t = 0$. In the next year 0.8, or 8000, of these newborns will have survived and will now be of age 1. They will produce 0.3×8000, or 2400, progeny. In year 2, 0.9 of the 8000, or 7200, 1-year-olds will survive, and 0.8 of the 2400, or 1920, newborns will survive and become 1-year-olds. The 1920 of age 1 will produce 0.3×1920, or 576, offspring, and the 7200 of age 2 will produce 0.4×7200, or 2880, offspring. The total number of offspring is 3456. You might want to carry this procedure forward for another year or two to check your understanding.

In Table 6-2 this process is continued for 25 years. The numbers in each age group are given as a proportion of the total rather than in absolute numbers.

TABLE 6-1

The change in composition of a population with a fixed schedule of age-specific survival and fertility values. The survival probabilities are $p_0 = 0.8$, $p_1 = 0.9$, $p_2 = 0.9$, $p_3 = 0.7$, $p_4 = 0.5$, $p_5 = 0$. The birth rates are $b_1 = 0.3$, $b_2 = 0.4$, $b_3 = 0.5$, $b_4 = 0.4$, $b_5 = 0$. Solid arrows indicate survival; dashed arrows, births.

	AGE						
TIME	0	1	2	3	4	5	TOTAL
0	10,000	0	0	0	0	0	10,000
1	2400	8000	0	0	0	0	10,400
2	3456	1920	7200	0	0	0	12,576

TABLE 6-2

The approach to a stable age distribution and the change in total numbers of the population from Table 6-1. After about 15 years the age structure has nearly stabilized and the population is growing at a constant rate of 1.37 percent per year.

TIME,	PROPORTION OF AGE						TOTAL NUMBER,	
t	0	1	2	3	4	5	N_t	$\frac{N_t}{N_{t-1}}$
0	1.0000	0	0	0	0	0	10,000	
1	0.2308	0.7692	0	0	0	0	10,400	1.0400
2	0.2748	0.1527	0.5725	0	0	0	12,576	1.2092
3	0.3026	0.1757	0.1098	0.4119	0	0	15,733	1.2511
4	0.2763	0.2225	0.1454	0.0909	0.2650	0	17,118	1.0880
5	0.2408	0.2243	0.2032	0.1328	0.0645	0.1344	16,869	0.9855
6	0.2755	0.1986	0.2082	0.1886	0.0958	0.0333	16,358	0.9697
7	0.2704	0.2098	0.1702	0.1784	0.1257	0.0456	17,185	1.0506
8	0.2635	0.2136	0.1864	0.1512	0.1233	0.0620	17,407	1.0129
9	0.2651	0.2099	0.1913	0.1670	0.1053	0.0613	17,488	1.0046
10	0.2681	0.2090	0.1861	0.1697	0.1152	0.0519	17,746	1.0148
12	0.2659	0.2105	0.1875	0.1645	0.1139	0.0577	18,268	1.0119
14	0.2665	0.2102	0.1863	0.1655	0.1150	0.0560	18,772	1.0143
16	0.2664	0.2102	0.1868	0.1653	0.1143	0.0565	19,284	1.0134
18	0.2664	0.2103	0.1866	0.1657	0.1145	0.0565	19,817	1.0137
20	0.2664	0.2102	0.1867	0.1658	0.1144	0.0564	20,361	1.0136
25	0.2664	0.2103	0.1867	0.1657	0.1145	0.0565	21,790	1.0137

During the first few years the population size varies widely and the age structure changes drastically. But something remarkable happens after about 15 years. The proportion in each age category has become nearly constant, while the population as a whole and the number in each age group are increasing by a factor 1.0137 each year. This example illustrates the age-stability property:

The population eventually attains a constant age composition, and the total population increases at a constant rate.

The time required to attain age stability varies with the pattern of births and deaths, but for populations that reproduce for a considerable length of time, such as that of the human, the time required to attain approximate equilibrium is about three times the maximum age of reproduction.

If you are familiar with matrix algebra, you will probably have noted that Equations 6-1 and 6-2 lend themselves to matrix form. If you would like to pursue this possibility, these equations are given in matrix notation in Appendix 5.

We are intersted in several properties of this schedule of age-specific birth and death rates, for example, the probability of survival to a specified age. The probability of a newborn still being alive at age x is designated l_x and is the product of surviving each of the previous time periods. Hence

$$l_x = p_0 p_1 p_2 \cdot \cdot \cdot p_{x-1}. \tag{6-3}$$

Notice that

$$l_{x+1} = l_x p_x. \tag{6-4}$$

The **life expectancy** is the total number of years we expect a newborn to live. The life expectancy at birth is given by

$$E_0 = l_1 + l_2 + l_3 + \dots + l_m = \sum_1^m l_x. \tag{6-5a}$$

Equation 6-5a assumes that deaths occur immediately after enumeration. If deaths occur uniformly throughout the year, the life expectancy is roughly half a year longer.

We can perhaps clarify the meaning of life expectancy if we define it as the mean age of death. In symbols, this definition is

$$E_0 = \sum_1^m x l_x q_x. \tag{6-5b}$$

But q_x, the probability of death, is

$$q_x = 1 - p_x = 1 - \frac{l_{x+1}}{l_x} = \frac{l_x - l_{x+1}}{l_x}.$$

Making this substitution, we have

$$\begin{aligned} E_0 &= \sum [x(l_x - l_{x+1})] = \sum xl_x - \sum xl_{x+1} \\ &= 1l_1 + 2l_2 + 3l_3 + \ . \ . \ . \ + ml_m \\ &\qquad - 1l_2 - 2l_3 - \ . \ . \ . \ - {}_{m-1}l_m \\ &= \sum_1^m l_x, \end{aligned}$$

which is identical to the previous definition.

In the example of Table 6-1,

$$\begin{aligned} l_1 &= p_0 = 0.800 \\ l_2 &= p_0 p_1 = 0.720 \\ l_3 &= p_0 p_1 p_2 = 0.648 \\ l_4 &= p_0 p_1 p_2 p_3 = 0.454 \\ l_5 &= p_0 p_1 p_2 p_3 p_4 = 0.227 \\ l_6 &= p_0 p_1 p_2 p_3 p_4 p_5 = 0 \end{aligned}$$

$$E_0 = 2.85 \text{ years.}$$

We are also interested in the rate of reproduction. For example, starting with a newborn female, what is the total number of daughters that she can be expected to produce? To calculate this number, we compute the probability of surviving to age x, multiply this value by the number of progeny produced at that age, and sum this product for all ages. This value is called the **net reproductive rate** and is conventionally designated as R_0. In symbols,

$$R_0 = l_1 b_1 + l_2 b_2 + \ . \ . \ . \ + l_k b_k = \sum_1^k l_x b_x. \qquad 6\text{-}6$$

In the numerical example in Table 6-2, $R_0 = 1.034$.

If R_0 is greater than 1, the population will eventually increase, although temporary decreases may occur because of nonequilibrium age structure. On the other hand, if R_0 is less than 1, the population will eventually decrease.

Another quantity of interest is the mean age of reproduction of a group of newborn females, which is given by

$$T = \frac{\sum_1^k x l_x b_x}{\sum_1^k l_x b_x} = \sum_1^k \frac{x l_x b_x}{R_0}. \qquad 6\text{-}7$$

In the numerical example, $T = 2.514/1.034 = 2.433$.

Remember from Table 6-2 that the rate of increase per year, after the population has attained age stability, is 1.0137. This value corresponds to a rate per generation of $1.0137^T = 1.034$, which is in agreement with the value of R_0, as expected.

Finally, we may want to know the future progeny expected from a female of a specified age, x. This quantity is called R_x and is given by

$$R_x = \frac{1}{l_x}[l_{x+1} b_{x+1} + l_{x+2} b_{x+2} + \ldots + l_k b_k]. \qquad 6\text{-}8$$

In the example $R_1 = 0.793/0.8 = 0.992$.

More about Age Stability. Although one numerical example hardly constitutes a proof, Table 6-2 illustrates a theorem that can be proved: except in special cases, an age-structured population with a fixed schedule of age-specific birth and death rates eventually attains a stable age distribution, and when this state is reached, each age group and the total population increase at a constant rate.

An example of a population that does not approach a constant age structure is one in which the birth rates at ages 1, 3, 5, 7, . . . , are all 0. This population will approach a state in which the population numbers are different in even- and odd-numbered years, and the age structure permanently oscillates.

This example, of course, is artificial and contrived. A sufficient condition for the population to attain a stable age distribution is that any two successive age classes have birth rates greater than 0. This condition is almost always met in organisms that reproduce more than once. However, in 17-year cicadas, for example, we might expect the numbers in different age groups to cycle. At the bottom of Table 6-2, we can see that the population increases each year by a factor 1.0137. This factor is conventionally designated by λ. Its value is inherent in the schedule of age-specific birth and death rates.

We can, of course, always determine the value of λ by grinding out the numbers in the population year by year until the rate of increase stabilizes, as was done in Table 6-2. We can also find the proportions in each age class in the age-stable population the same way. But let us find a better way.

To find the stable age distribution, we note that when age stability is reached, the number in each class increases by a factor λ each year. Thus

$$n_{x,t} = n_{x,t-1}\lambda = n_{x,t-2}\lambda^2 = n_{x,t-i}\lambda^i.$$

In particular, the number of newborns i years ago is

$$n_{0,t-i} = n_{0,t}\lambda^{-i}.$$

Thus the number of individuals of age x at time t is equal to the number born x years ago times the probability of surviving to age x. In symbols,

$$n_{xt} = n_{0,t-x}\,l_x = n_{0t}\lambda^{-x}l_x. \qquad \text{6-9}$$

Using Formula 6-9, we can compute the proportion in any age class, provided we know λ.

Now, how do we compute λ? From Equation 6-2 we have

$$n_{0t} = n_{1t}b_1 + n_{2t}b_2 + \ . \ . \ . \ + n_{kt}b_k = \sum_{x=1}^{k} n_{xt}b_x.$$

Now we substitute from Equation 6-9, leading to

$$\begin{aligned} n_{0t} &= n_{0t}l_1\lambda^{-1}b_1 + n_{0t}l_2\lambda^{-2}b_2 + \ . \ . \ . \ + n_{0t}l_k\lambda^{-k}b_k \\ &= n_{0t}\sum_{x=1}^{m} l_x\lambda^{-x}b_x. \end{aligned}$$

Canceling n_{0t} from both sides and multiplying by λ^k, we have

$$\lambda^k - l_1b_1\lambda^{k-1} - l_2b_2\lambda^{k-2} - \ . \ . \ . \ - l_kb_k = 0. \qquad \text{6-10}$$

This equation has only one change of sign, hence only one positive root. (If you have forgotten this rule, it is Descartes' rule of signs.) We also know that unless the population is exploding or rapidly going extinct, λ cannot be very far from 1. So using 1 as a trial value, we can easily find the root by trial and error. Of course, we can use any of the standard methods for obtaining a numerical solution to a polynomial equation.

Returning to the now familiar numerical example in Table 6-2, we find from Equation 6-9 that the equilibrium ratios in the different age classes are

$$\begin{aligned} n_0 &= 1.000n_0 \\ n_1 &= 0.7892n_0 \qquad (= l_1\lambda^{-1}n_0) \\ n_2 &= 0.7007n_0 \qquad (= l_2\lambda^{-2}n_0) \\ n_3 &= 0.6221n_0 \qquad (= l_3\lambda^{-3}n_0) \\ n_4 &= 0.4300n_0 \\ n_5 &= 0.2121n_0. \end{aligned}$$

Adding these numbers, the total population, N, is

$$N = 3.7541 n_0 \qquad \left[= \left(\sum_{0}^{m} l_x \lambda^{-x} \right) n_0 \right].$$

Dividing by the total, N, gives 0.2664, 0.2101, 0.1866, 0.1657, 0.1145, and 0.0565 for the proportions in the six age classes, as the bottom line of Table 6-2 indicates.

The equation for λ is

$$\lambda^4 - 0.240\lambda^3 - 0.288\lambda^2 - 0.324\lambda - 0.182 = 0,$$

with the positive root 1.0137, as we obtained in Table 6-2 by carrying the population forward for 25 years.

Since the birth rate is the number of newborns divided by the total population, we can use the preceding expressions to find what the birth rate will be when the

TABLE 6-3

An illustration of the death rates at different ages in the life cycle of a gall insect, *Urophora jaceana*.

MONTH	DENSITY/METER²	CAUSE OF DEATH	DEATH RATE, q_x	SURVIVAL, p_x	CUMULATIVE SURVIVAL, l_x
July	203.0				
		Infertile eggs	0.090	0.910	0.910
	184.7				
		Failure to form gall	0.201	0.799	0.727
	147.6				
		Died in gall	0.020	0.980	0.713
	144.6				
		Eurytoma curta	0.455	0.545	0.388
August	78.8				
		Parasites	0.365	0.635	0.247
	50.0				
		Probably mice	0.616	0.384	0.095
Winter	19.2				
		Mice	0.635	0.365	0.035
	7.0				
		Unknown	0.257	0.743	0.026
	5.2				
		Birds and parasites	0.308	0.692	0.018
May	3.6				
		Floods	0.436	0.564	0.010
July	2.03				
Total					0.010

Data from G. C. Varley. 1947. *J. Anim. Ecol.* 16:139.

population is at age stability. Designating this value by $\bar{b}$, we have

$$\bar{b} = \frac{n_0}{N} = \frac{1}{\sum_0^m l_x \lambda^{-x}}. \qquad \text{6-11}$$

This theory strictly applies only to a population in which all the individuals reproduce at the same time and that is censused immediately after. If the births are spread over a longer period, it will still yield reasonably accurate results as long as the censuses are frequent relative to the whole life cycle. For example, this theory would be quite accurate for the human population if the population were censused every year. For strict accuracy we must account for newborns that die before being censused, deaths at various times during the year, and so on. Demographers have various formulae for converting the discrete data into smooth curves. These details are what make actuarial science difficult, but here we are concerned only with a general understanding.

The data in Table 6-3 illustrate how mortality acts at various ages in an insect species. The total mortality of these gall flies is about 99 percent, compensated, of course, by a correspondingly high fertility.

6-2 CONTINUOUS MODEL

Since most populations reproduce at many times and deaths occur continuously through the life span, it is more natural to treat the process by continous functions, using the methods of the calculus. The equations are the continous analogs of those we have been using.

If the population is large, the discreteness introduced by individual births and deaths is lost in the large total, and we can regard the process of population change as essentially continuous. Mathematically we think of the process as the limit of the previous method as the age intervals get smaller.

The equations for dealing with such a population were developed by Lotka and Fisher (summarized in Lotka, 1956, and Fisher, 1958). We use Fisher's procedure here. To be concrete, I shall again think of a human population with time measured in years.

Let $l(x)$ be the probability of survival from birth to age x. It is no longer necessary for x to take only integral values; at some instant a person may be 27.54678 years old. Let the probability of reproducing during the infinitesimal age interval from x to $x + dx$ be $b(x)\,dx$. Then the probability of a newborn surviving to age x and reproducing during the next time interval dx is $l(x)b(x)\,dx$. The expected number of daughters per newborn female during her whole lifetime is this quantity summed over all ages, or

$$R(0) = \int_0^A l(x)b(x)\,dx, \qquad 6\text{-}12$$

where A is the highest age at which reproduction is possible. For mathematical convenience we replace A with ∞, which does not change anything, since $b(x) = 0$ for all ages beyond A. This equation is the continuous equivalent of Equation 6-6. If $R(0)$ is greater than 1, the population will eventually increase, although temporary decreases may occur because of changes in the age distribution. If $R(0)$ is less than 1, the population will eventually decrease and ultimately become extinct.

In the previous section we saw that a population with a constant set of age-specific birth and death rates attains a stable age distribution. In this state the population increases or decreases at a constant rate. To correspond to λ in the discrete case, we use e^m in the continuous model, where m is the **Malthusian parameter.** We shall now see how we can determine m from the birth and death rates.

Of the population now alive (at time t, say), those of age x were born x years ago at time $t - x$. Let the total number of births occurring in the interval t to $t + dt$ be $B(t)\,dt$. If this rate were continued for a year, there would be $B(t)$ births. The number of births x years ago was $B(t - x)$. Of those born at that time, a fraction $l(x)$ will still be alive, and of this group a fraction $b(x)\,dx$ will give birth during the interval dx. Thus the current birth rate of persons of age x will be $B(t - x)l(x)b(x)$. This value, summed over all ages, is the total birth rate at time t, which gives us

$$B(t) = \int_0^\infty B(t - x)l(x)b(x)\,dx.$$

If the population has attained age stability, its size and birth rate are increasing at a rate m. That is, the equation for the rate of change of the whole population number is

$$\frac{dN}{dt} = mN,$$

which integrates to

$$N(t) = N(0)e^{mt}.$$

In x years the population size and the number of newborns will have increased by a factor e^{mx}. Therefore the number of newborns x years ago was a fraction e^{-mx} of the current number. That is,

$$B(t-x) = B(t)e^{-mx}.$$

Substituting this expression into the expression of $B(t)$ and canceling $B(t)$ from the two sides of the equation yields

$$1 = \int_0^\infty e^{-mx} l(x) b(x)\, dx. \qquad \text{6-13}$$

This equation provides a means for calculating m if the functions $1(x)$ and $b(x)$ are known.

Equation 6-13 is analogous to Equation 6-10. Perhaps the analogy will be more apparent if we divide by λ^k and rewrite the equation as

$$1 = \sum_{x=0}^{\infty} \lambda^{-x} l_x b_x. \qquad \text{6-14}$$

λ corresponds to e^m. Just as Equation 6-10 has only one positive real root, Equation 6-14 has only one real solution for m. We can readily find this solution by trial and error, starting with a trial solution and adjusting m until the right and left sides of the equation are equal.

When the population has reached age stability, the rate of change of the total number is given by

$$\frac{dN(t)}{dt} = mN(t). \qquad \text{6-15}a$$

In integrated form, letting $N(0)$ be the initial number (at some time after age stability has been attained) and $N(t)$ the number t years later, we have

$$N(t) = N(0)e^{mt}. \qquad \text{6-15}b$$

Table 6-4 gives the Malthusian parameters corresponding to the birth and death rates in the United States in recent years. As mentioned earlier, these parameters give the rate at which the population would increase if it were at age equilibrium under the current schedule of age-specific birth and death rates. The changes in the Malthusian parameter as a result of changing death and, more important, birth rates in recent years show strikingly in the data. At the 1940 rates the population would have been nearly stable. Then substantial increases occurred in the fertility rates, followed by a reduction in recent years. If current rates continue, the population will eventually decrease, as indicated by the negative sign of m. (Of course, I have neglected immigration, which in the United States is an important part of the population picture.)

TABLE 6-4

The Malthusian parameters corresponding to birth and death rates in the United States population for specific years. These are the rates at which the population would eventually decrease or increase if the birth and death rates persisted. The actual rate of increase is in the right column.

DATE	MALTHUSIAN PARAMETER	ACTUAL RATE OF INCREASE
1935–1939	−0.0018	0.0085
1946–1949	0.0135	0.0157
1950–1954	0.0168	0.0159
1955–1959	0.0211	0.0159
1960–1964	0.0186	0.0134
1965–1969	0.0082	0.0094
1970–1974	0.0000	0.0073
1975	−0.0059	0.0063
1976	−0.0065	0.0058
1977	−0.0052	0.0066
1978	−0.0057	0.0065

In practice we must use discrete formulae because of the way census data are recorded. The accuracy of the calculations can be improved by suitable conventions regarding the way birth and death rates are computed (for example, yearly births divided by the appropriate age group counted at the center of the age period) and by methods of smoothing the discrete data to get a continuous curve. Such methods are outside the scope of this book.

Table 6-5 compares the formulae for discrete and continuous models. Their correspondence is apparent. I have used symbols such as b_x for discrete generation models and $b(x)$ for continuous models to emphasize that the latter are continuous functions.

Two of the comparisons call for comment. Line 10 gives the birth rate (ratio of newborns to total) when the population has a stable age distribution. This rate may differ considerably from the actual ratio if the population is not at equilibrium. Line 4 gives the mean age of reproduction of a group of females born at the same time and followed through their lifetimes. Line 9 gives the mean age of mothers now giving birth in an equilibrium population. The two are not the same unless the Malthusian parameter is 0. If the population is growing, line 9 has a smaller value than line 4 because in a growing population more individuals are in the younger ages. Line 9 gives the mean generation length if we wish to convert from years to generations as the unit of time measurement—for example, if we want to use the discrete generation formulae of Chapters 1 through 5 as approximations to a continuous model.

TABLE 6-5

Summary of formulae for discrete and continuous models.

	DISCRETE	CONTINUOUS
Properties of an individual or cohort		
1. Probability of survival to age x (and beyond)	$p_0 p_1 \cdots p_{x-1} = l_x$	$l(x)$
2. Life expectancy (E_0)	$\sum_1^\infty l_x$	$\int_0^\infty l(x)\,dx$
3. Expected lifetime offspring of a newborn (R_0)	$\sum_1^\infty l_x b_x$	$\int_0^\infty l(x)b(x)\,dx$
4. Mean age of reproduction (T)	$\dfrac{\sum_1^\infty l_x b_x x}{R_0}$	$\dfrac{\int_0^\infty l(x)b(x)x\,dx}{R(0)}$
5. Expected future offspring at age x, R_x, or $R(x)$	$\dfrac{1}{l_{(x)}} \sum_{y=x+1}^\infty l_y b_y$	$\dfrac{1}{l(x)} \int_x^\infty l(y)b(y)\,dy$
Properties of an equilibrium population		
6. Definition of λ and m	$\sum_x \lambda^{-x} l_x b_x = 1$	$\int_0^\infty e^{-mx} l(x)b(x)\,dx = 1$
7. Age distribution	$n_x = n_0 l_x \lambda^{-x}$	$n(x) = n(0)l(x)e^{-mx}$
8. Population growth rate and size	$N_t = \lambda N_{t-1} = \lambda^t N_0$	$\dfrac{dN}{dt} = N_t m, \quad N_t = N_0 e^{mt}$
9. Mean age of reproduction, $\bar{x}_r$	$\sum_0^\infty \lambda^{-x} l_x b_x x$	$\int_0^\infty e^{-mx} l(x)b(x)x\,dx$
10. Mean birth rate, $\bar{b}$	$\dfrac{1}{\bar{b}} = \sum_0^\infty \lambda^{-x} l_x$	$\dfrac{1}{\bar{b}} = \int_0^\infty l(x)e^{-mx}\,dx$
Reproductive value		
11. Reproductive value at age x relative to a newborn, v_x/v_0	$\dfrac{\lambda^x}{l_x} \sum_{y=x+1}^\infty \lambda^{-y} l_y b_y$	$\dfrac{e^{mx}}{l(x)} \int_x^\infty e^{-my} l(y)b(y)\,dy$
12. Reproductive value of a newborn, v_0 or $v(0)$	$v_0 = \dfrac{1}{\bar{b}\bar{x}_r}$	$v(0) = \dfrac{1}{\bar{b}\bar{x}_r}$

6-3 REPRODUCTIVE VALUE

In his classic book, *The Genetical Theory of Natural Selection* (1930, 1958), R. A. Fisher asked the genetically relevant question: To what extent does an individual of age x contribute to the ancestry of future generations? To answer this question, he defined the quantity $v(x)$ as the **reproductive value** of an individual of age x.

Obviously the reproductive value is 0 for an individual that is past the reproductive age. It is clearly less at birth than a few years later, since a person of age 10, say, has a better chance of surviving to reproduce than a newborn infant. Furthermore the person's reproduction will begin sooner, which in a growing population will increase the total contribution to future generations because of an earlier start. We might expect the value to be maximum somewhere near the beginning of the reproductive period.

The contribution of an age group to future generations depends mainly on how long individuals of this age survive and the number of births at each age during the survival period. This contribution will be the probability of surviving until a certain age x multiplied by the expectation of progeny at that age, weighted by a factor that discounts progeny born in the distant future, much as the present value of a loan to be repaid in the future is discounted. (In this formula we assume that the population is growing; if it is contracting, the weight is greater for distant progeny—equivalent to negative interest rates, if there is such a thing.)

If we put these relations together, the reproductive value at age x is

$$\frac{v(x)}{v(0)} = \int_x^\infty l(x,y)b(y)w(y-x)\,dy, \tag{6-16}$$

where $l(x,y)$ is the probability of surviving from age x to age y (and beyond), $w(y-x)$ is a weighting factor, and $v(0)$ is the reproductive value of a newborn.

To decide on an appropriate weighting factor, consider two cohorts born at times t_0 and $t_0 + k$. After a few generations we reach a stage in which the descendants of each cohort will be growing exponentially. As we saw from Table 6-2, after about three generations the two sets of descendants will be growing at the same rate and will be in a constant ratio. The growth pattern followed by each cohort will be exactly the same. Figure 6-2 illustrates this phenomenon.

At time $t_1 + k$ the ratio of N_1, the descendants of the second cohort, to N_1', the descendants of the first, will be e^{-mk} because at this time the two groups are each increasing (or decreasing) exponentially at the same rate m. We choose $w(y-x)$ to be the eventual contribution of the descendants of an age group born at time y relative to those born at time x. Since $y - x = k$, we replace $w(y-x)$ by $e^{-(y-x)m}$ in Equation 6-16.

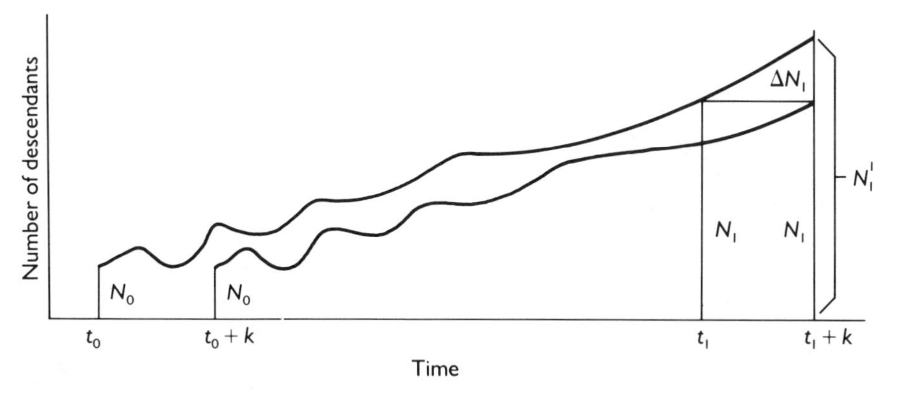

FIGURE 6-2. The number of descendants of two groups born at times t_0 and $t_0 + k$ at various times in the future. Both numbers follow complicated, but identical, patterns, which finally smooth out as the descendants of each group reach a stable age distribution.

Furthermore $l(x,y) = l(y)/l(x)$. Making these substitutions into Equation 6-16 and rearranging, we obtain

$$\frac{v(x)}{v(0)} = \frac{e^{mx}}{l(x)} \int_x^\infty l(y)b(y)e^{-my}\,dy. \qquad 6\text{-}17$$

This equation is Fisher's definition of reproductive value. Notice that when $x = 0$, then $e^{mx} = 1$, $l(x) = 1$, and $v(x) = v(0)$, as they should (see Equation 6-13).

Definition of $v(0)$. How should we define $v(0)$? Fisher simply ignored this question and treated only the ratio of $v(x)$ to $v(0)$, but we can do better. Let us define $v(0)$ in such a way that after age stability has been reached, the total reproductive value is the same as the total population number. In that case, letting N stand for the total population number and V for the total reproductive value of the population, we have

$$N = V = \int_0^\infty n(x)v(x)\,dx.$$

Substituting for $v(x)$ from Equation 6-17 gives

$$N = \int_0^\infty \frac{n(x)\,v(0)e^{mx}}{l(x)}\,dx \int_x^\infty e^{-my}l(y)b(y)\,dy.$$

However, when age stability is reached, $n(x) = n(0)l(x)e^{-mx}$ (from Table 6-5, line 7).

Thus

$$\begin{aligned} N &= n(0)v(0) \int_0^\infty \int_x^\infty e^{-my} l(y) b(y)\, dy\, dx \\ &= n(0)v(0) \int_0^\infty x l(x) b(x) e^{-mx}\, dx \\ &= n(0)v(0)\, \bar{x}_r. \end{aligned}$$

The last substitution comes from line 9 in Table 6-5. But at age stability, $n(0)/N$ is the number of newborns divided by the total population number, which is the population birth rate $\bar{b}$. Substituting this expression at last gives us what we are after — a definition of $v(0)$:

$$v(0) = \frac{1}{\bar{x}_r \bar{b}}. \qquad \text{6-18}$$

The exponential term in Equation 6-17 diminishes the value of progeny born in the future. To use the money analogy again, money to be received in the future is discounted by an amount determined by prevailing interest rates. If the population is decreasing, the present value is inflated rather than discounted. If the age-stable population is of constant size, then $m = 0$ and the exponential term disappears.

Putting all this information together, we can write an equation for the total population number in the future, giving each individual in the present population a weight determined by the reproductive value for that individual's age. The formula is

$$\begin{aligned} N(t) &= V(0)e^{mt} \\ V(0) &= \int_0^\infty n(x,0)v(x,0)\, dx. \end{aligned} \qquad \text{6-19}$$

Figure 6-3 shows that this procedure works. The actual population at time t, $N(t)$, and the total reproductive value at that time, $V(t)$, are both plotted. We can see that whereas the total population number jumps around capriciously, the reproductive value increases smoothly and eventually coincides with the actual population number when age stability is reached. The data are from Table 6-1.

To summarize, the reproductive value has the following two properties:

1. It assigns a weight to each age class in proportion to the contribution of an individual of that age to the population in the future after the age irregularities have ironed out (usually about three generations).

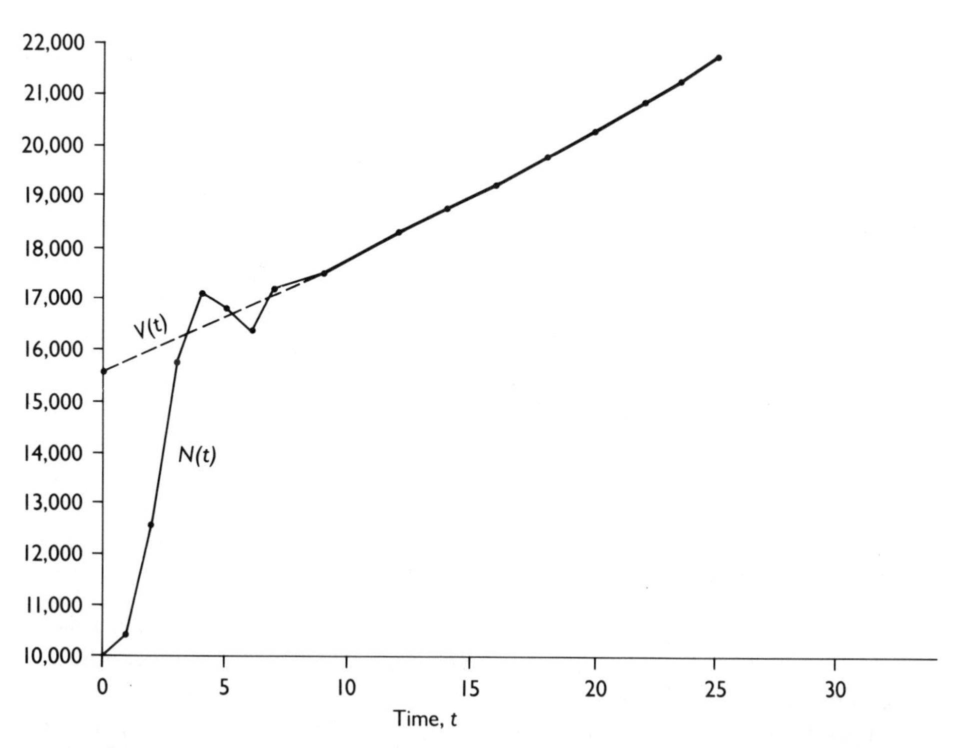

FIGURE 6-3. The total population number, $N(t)$, and total reproductive value, $V(t)$, at various times starting with 10,000 newborns. The survival and fertility parameters are those of Table 6-1. Although the actual population size varies erratically in the early time periods, the total reproductive value changes uniformly.

2. If each age class is weighted by its reproductive value, the total weighted number increases at a constant rate, the same as that at age stability, and eventually this number coincides with the actual population number.

We shall now see how to calculate the numbers plotted in Figure 6-3. Since the data are in discrete intervals, we need discrete generation formulae, which are in Table 6-5. The actual values are

$$b = 0.26640 \qquad \bar{x}_r = 2.41803 \qquad v(0) = 1.55238$$
$$v(1) = 1.5013 \qquad v(2) = 1.06909 \qquad v(3) = 0.42881$$
$$\lambda = 1.013655 \qquad \lambda^{-1} = 0.986529 \qquad v(4) = 0.$$

Using these values, wc can compute the reproductive values of each group, given in Table 6-6, for the data in Table 6-1. The right column gives the ratio of the total reproductive value in a generation to that in the previous generation; this ratio

TABLE 6-6

Reproductive values for successive generations in the hypothetical population from Table 6-1.

	REPRODUCTIVE VALUE OF THE GROUP OF AGE							
TIME, t	0	1	2	3	4	5	TOTAL VALUE	$\frac{V_t}{V_{t-1}}$
0	15,524	0	0	0	0	0	15,524	
1	3726	12,010	0	0	0	0	15,736	1.0136
2	5365	2883	7703	0	0	0	15,951	1.0136
3	7391	4150	1848	2779	0	0	16,168	1.0136
4	7342	5718	2663	667	0	0	16,390	1.0137
.	.	.	.	.	.	.	.	.
.	.	.	.	.	.	.	.	.
.	.	.	.	.	.	.	.	.
25	9013	6879	4353	1548	0	0	21,793	1.0137

immediately takes the value 1.0136, which is the ultimate rate of population growth. The total reproductive value is sometimes called the **stable equivalent population** (see, for example, Keyfitz, 1968).

For example, the United States population in 1964 was about 191 million. The stable equivalent population (total reproductive value) was 177 million. Clearly the population is not in a stable age distribution. If we wanted to extrapolate 75 or more years into the future using the Malthusian parameter, we would do better to start with 177 million rather than 191 million. For contrast, the population of Hungary that year was 10.1 million, and the total reproductive value was 12.9 million; the population is out of equilibrium the other way. A larger fraction of the population is in the reproductive ages in Hungary than in the United States. The population of Chile was 8.4 million, and V was 8.5 million. These values do not necessarily mean that the population is at age equilibrium; departures at different ages may cancel. But the 1964 population of Chile is a better guide to the future than that of either the United States or Hungary.

Table 6-7 gives some reproductive values for various human populations in 1964. The values given are relative to the values at birth, that is, $v(x)/v(0)$. Notice that in populations with a high childhood mortality and growth rate, the reproductive value in the late teens is considerably higher than it is at birth. There are two reasons for this phenomenon. The major reason is that because of a high childhood mortality, a teenager is much more likely to survive to reproduce than an infant is. The second reason is that the children produced by a teenager are born sooner than those of a newborn and hence are discounted less.

Note that in the United States the female reproductive value peaks earlier than that of the male because of the earlier age of reproduction (although this comparison is

TABLE 6-7

Reproductive values of some representative human populations. The values are those relative to a newborn by 5-year periods.

	CHILE, 1964	GERMANY, 1964	UNITED STATES, 1964	
AGE	FEMALES	FEMALES	FEMALES	MALES
0–4	1.186	1.040	1.063	1.075
5–9	1.347	1.076	1.154	1.178
10–14	1.503	1.111	1.250	1.289
15–19	1.581	1.121	1.261	1.391
20–24	1.406	0.975	0.997	1.276
25–29	1.029	0.638	0.575	0.884
30–34	0.633	0.305	0.264	0.487
35–39	0.297	0.108	0.091	0.230
40–44	0.088	0.025	0.018	0.094
45–49	0.014	0.002	0.001	0.035
50–54	0.002	0.000	0.000	0.010

Data from Keyfitz. 1968.

somewhat obscured by the 5-year grouping). However, the male value persists longer because of the longer male reproductive period.

How does all this information relate to Mendelian populations? As you can see from the difficulties we have been having when Mendelism is not taken into account, dealing with the reproductive value of a Mendelian population is not easy. Actually research in this area is in its infancy. My investigations (see Crow, 1979) seem to show that reproductive value weighting works reasonably well, although not perfectly, for tracing allele frequencies and for the rate of change of population fitness. Much work remains to be done. If you want to pursue this topic further, the standard reference is *Evolution in Age-structured Populations* by B. Charlesworth (1980).

Now, for a bit of an anticlimax. We have considered these elegant properties of age structure and reproductive value. Yet for most problems in evolution and population genetics, we can ignore such niceties. Ordinarily we are interested in trends over many generations, during which the age distribution has plenty of time to equilibrate. If we want to ask about allele frequency changes, assuming constant fitness parameters, then we can assume that the population is in age equilibrium unless selection is so strong as to appreciably change the composition in three or four generations. For a rigorous discussion of this topic, see Nagylaki (1977).

In the next section we assume that the population is in a stable age distribution, so that its rate of increase is measured by the Malthusian parameter, and we consider the effects of crowding on the rate of growth.

6-4 REGULATION OF POPULATION NUMBER

Nothing is more obvious than the fact that exponential growth of a population cannot continue forever. Sooner or later something becomes limiting. The environment will only sustain so many organisms, or territorial behavior limits the amount of available space, or some other factor limits further growth. We hope that the human population will be regulated by reduction of birth rates rather than by the tragedy of higher death rates.

In this section we shall consider some of the classical equations used by ecologists to take population limitation into account. Assuming age stability, we can modify Equation 6-15 to read

$$\frac{dN}{dt} = aN - f(N), \tag{6-20}$$

where $f(N)$ expresses some change in the rate of increase, so that the rate is diminished as the population gets larger. I have written $N(t)$ simply as N, but remember that it is still a function of t.

One of the simplest assumptions is to let $f(N)$ be proportional to N^2. Then we can write Equation 6-20 as

$$\frac{dN}{dt} = aN - bN^2. \tag{6-21}$$

In this equation a represents the growth rate if the population were uncrowded; b represents the effect of population number in diminishing the growth rate. We can give this equation a concrete interpretation by writing it as

$$\frac{dN}{dt} = rN\left(\frac{K-N}{K}\right), \tag{6-22}$$

where $a = r$ and $b = r/K$. Then r is the **intrinsic rate of increase,** the rate at which the population would grow if there were no crowding effect, and K is the **carrying capacity** of the environment for this species. When $N = K$, the population ceases to grow. If the population exceeds K, as might happen if the carrying capacity is reduced (by a drought, for example) for a population already at equilibrium size, the growth dN/dt will be negative.

We can rewrite Equation 6-22 as

$$\frac{dN}{N} + \frac{dN}{K-N} = r\,dt,$$

which is readily integrated to give

$$t = \frac{1}{r} \ln \frac{N_t(K - N_0)}{(K - N_t)N_0}. \qquad 6\text{-}23$$

We have encountered this equation before—notice its similarity to Equation 4-8.

For example, if the intrinsic rate of increase of a population is 1 percent per year ($r = 0.01$) and the carrying capacity, K, is 5000, then the time required to double the population's number from $N_0 = 1000$ to $N_t = 2000$ is

$$t = \frac{1}{0.01} \ln \frac{2000 \times 4000}{3000 \times 1000} = 98.1 \text{ years.}$$

If the rate were not regulated, the time required (see Equation 6-15*b*) would be

$$t = \frac{1}{r} \ln \frac{N_t}{N_0} = 69.3 \text{ years.}$$

Notice that, whether the growth rate is regulated or not, the time required for a specified change is proportional to $1/r$. For example, if r were 10 times as small, or 0.001, the times required for the regulated and nonregulated populations would be 10 times as large, or 981 and 693 years, respectively. This principle is similar to that for the time required to change allele frequencies by a specified amount with slow selection (see Section 4-2).

We can also write Equation 6-23 in the inverse form, giving the population number at time t as a function of t and the initial number N_0:

$$N_t = \frac{K}{1 + C_0 e^{-rt}}, \qquad C_0 = \frac{K - N_0}{N_0}. \qquad 6\text{-}24$$

Figure 6-4 shows this function graphically. It is symmetrical about the value $N = K/2$, at which time the increase in population number is maximum, and approaches K as a limit. This equation is often called the logistic curve and is widely used in classical ecology. Of course, it is only one of many equations that can be formulated. The logistic curve owes its popularity to its symmetry, simplicity, and its ease of interpretation. In many cases it provides a satisfactory approximation to reality.

Another equation that is sometimes used is the Gompertz curve:

$$\frac{dN}{dt} = rN\left(\frac{\ln K - \ln N}{\ln K}\right). \qquad 6\text{-}25$$

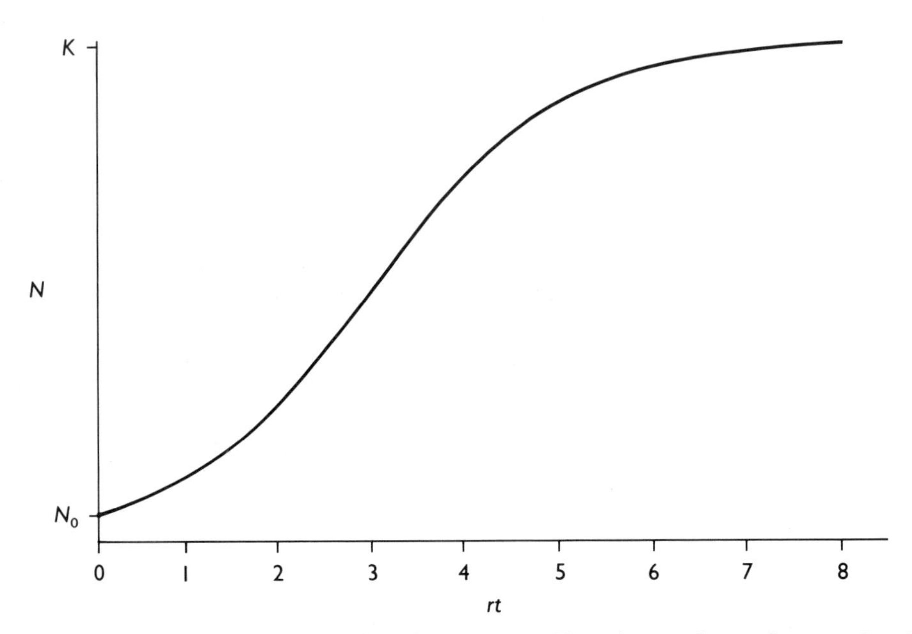

FIGURE 6-4. The logistic curve of population increase. The ordinate is the population number, N. The abscissa is rt, the product of the intrinsic rate of increase and the time. N_0, the initial population number, is taken as $1/20$ of the value of K, the carrying capacity.

This equation differs from the logistic curve in being regulated by the difference between the logs of K and N rather than the difference between K and N. It leads to a curve that is not symmetrical about the value $K/2$ and it fits some data better than the logistic curve does.

Of course, by using more complicated equations with more than two parameters, we can more closely fit any set of data. Such equations are often useful for description and prediction, especially for applied problems, but usually give little insight into the mechanisms.

6-5 COMPETING GROUPS

Suppose that we have two groups in the same environment and wish to determine how they change in relative frequency. These groups may be species, subspecies, or asexual strains. The only requirement is that they do not interbreed. I shall designate the two groups by subscripts 1 and 2. N_1 and N_2 denote the numbers in the two groups, and N is the total of the two.

Assume that the two groups are both at age equilibrium and let their Malthusian parameters be m_1 and m_2. Using Equation 6-15, we can then write,

$$\frac{1}{N_1}\frac{dN_1}{dt} = \frac{d \ln N_1}{dt} = m_1$$
$$\frac{1}{N_2}\frac{dN_2}{dt} = \frac{d \ln N_2}{dt} = m_2. \qquad 6\text{-}26$$

Now let

$$z = \ln \frac{N_1}{N_2} = \ln N_1 - \ln N_2. \qquad 6\text{-}27$$

From this definition and Equations 6-26, we find

$$\frac{dz}{dt} = m_1 - m_2 = s. \qquad 6\text{-}28$$

In general m_1 and m_2, and therefore s, are not constants but are functions of the degree of crowding, determined mainly by the total number N. However, if the population is near its carrying capacity, s may be nearly constant. We shall think of s as $s(N)$, although I shall suppress the N in the following formulae.

Now let p and $q = 1 - p$ be the proportions of the two groups. Then $N_1/N_2 = p/(1-p)$ and $z = \ln p - \ln(1-p)$. Using these equations, we find

$$\frac{d}{dt}[\ln p - \ln(1-p)] = \frac{dp}{dt}\left[\frac{1}{p} + \frac{1}{1-p}\right] = s$$

and, rearranging, we have

$$\frac{dp}{dt} = sp(1-p). \qquad 6\text{-}29$$

If we compare this equation with Equation 4-6, we find that they are in the same form.

Mendelian Populations. We can extend the preceding analysis to Mendelian populations. We assume that the population is at equilibrium for age structure. In other words any change in allele frequencies is slow relative to the time required to adjust the age structure. Each genotype has a Malthusian parameter, which measures the rate at which its genes are increasing. We must keep in mind that the genotype is not transmitted as a unit; each individual contributes a random half of its genes. Table 6-8 gives a model.

TABLE 6-8

A model for a Mendelian population with continuous age structure. The population is assumed to be at age equilibrium.

GENOTYPE	A_1A_1	A_1A_2	A_2A_2
Frequency	P_{11}	$2P_{12}$	P_{22}
Malthusian parameter	m_{11}	m_{12}	m_{22}

Let N stand for the total number of genes, which is twice the number of individuals in a diploid population. N_1 and N_2 are the numbers of alleles A_1 and A_2. The proportions of the two alleles are p_1 and p_2; thus, for example, $p_1 = N_1/N$. The rate of increase of A_1 alleles is

$$\frac{dN_1}{dt} = NP_{11}m_{11} + NP_{12}m_{12}. \tag{6-30}$$

We can rewrite this equation as

$$\frac{1}{N_1}\frac{dN_1}{dt} = \frac{d \ln N_1}{dt} = \frac{N(P_{11}m_{11} + P_{12}m_{12})}{N_1}.$$

But since $p_1 = N_1/N$,

$$\frac{d \ln N_1}{dt} = \frac{P_{11}m_{11} + P_{12}m_{12}}{p_1}. \tag{6-31a}$$

Likewise

$$\frac{d \ln N_2}{dt} = \frac{P_{12}m_{12} + P_{22}m_{22}}{p_2}. \tag{6-31b}$$

Now define z as in Equation 6-27. Hence we have

$$\frac{dz}{dt} = \frac{d \ln N_1}{dt} - \frac{d \ln N_2}{dt}. \tag{6-32}$$

Since $z = \ln(p_1/p_2)$,

$$\frac{dz}{dt} = \frac{d}{dt}(\ln p_1 - \ln p_2) = \frac{dp_1}{dt}\frac{1}{p_1 p_2}. \tag{6-33}$$

Now putting Equations 6-31, 6-32, and 6-33 together, we find

$$\frac{dp_1}{dt} = p_1 p_2 \left[\frac{P_{11}m_{11} + P_{12}m_{12}}{p_1} - \frac{P_{12}m_{12} + P_{22}m_{22}}{p_2} \right]. \tag{6-34}$$

Now we are assuming that genotypic selection is slow enough not to upset the age-structure equilibrium. If our assumption is true, the selection is weak enough that, with random mating, the genotypes are close to Hardy-Weinberg ratios. Replacing P_{11}, P_{12}, and P_{22} with p_1^2, $p_1 p_2$, and p_2^2, we finally arrive at

$$\frac{dp_1}{dt} = p_1 p_2 [(m_{11} - m_{12})p_1 + (m_{12} - m_{22})p_2]. \tag{6-35}$$

If you compare this equation to Equation 4-4, you will see that they are essentially the same. Since we are assuming slow selection, $\bar{w}$ is very close to 1. Therefore we come to the following conclusion:

> **With weak selection the continuous and discrete models lead to essentially the same results.**

The m's represent absolute fitnesses. Since we are interested in the changes in the proportions of different genotypes rather than their absolute frequencies, we might as well express the fitnesses in relative terms. To do this, let $m_{11} = 0$, $m_{12} = -hs$, and $m_{22} = -s$. Also, let $p_1 = p$, and $p_2 = q$. Making these substitutions in Equation 6-35 leads to

$$\frac{dp}{dt} = spq[q + h(p - q)]. \tag{6-36}$$

This formula corresponds to Equation 4-5a for a discrete model.

What do we mean by weak selection? Essentially we mean that the selection is weak enough that gene frequency changes very little during the time required for the population to equilibrate in age structure. This time is two to three generations. Certainly selection differences of the order of 1 percent are acceptable. The equations are useful as approximations even when the differences are 5 percent or greater. For a rigorous assessment of the accuracy of continuous approximations, refer to Nagylaki (1977).

The equations of this section are derived more generally for multiple alleles and for two loci in C&K, pages 190 ff, and with greater rigor by Nagylaki and Crow (1974) and Nagylaki (1977).

In this chapter I have mainly emphasized demography and have not attempted to deal with the ecological conditions causing population limitation. This field of research is now very active. An up-to-date report on this work can be found in *Theory of Population Genetics and Evolutionary Ecology: An Introduction* by Roughgarden (1980). Finally, for a charming and erudite historical account by the dean of population ecologists, see *An Introduction to Population Ecology* by G. E. Hutchison (1978).

QUESTIONS AND PROBLEMS

1. A population of wolves rears one litter per year, and the parameters are as follows:

$$
\begin{array}{llll}
n_{00} = 120 & p_0 = 5/6 & l_0 = 1 & b_0 = 0 \\
n_{10} = \ \ 50 & p_1 = 3/5 & l_1 = 5/6 & b_1 = 3/5 \\
n_{20} = \ \ 30 & p_2 = 2/3 & l_2 = 1/2 & b_2 = 2/3 \\
n_{30} = \ \ \ \ 0 & p_3 = 0 & l_3 = 1/3 & b_3 = 1/2.
\end{array}
$$

(a) What will be the numbers at each age 1 year later? (b) What is the expected total offspring per individual born? (c) What is the mean age of reproduction of a cohort of newborns? (d) What is the equilibrium rate of increase, λ? (e) What is the equilibrium age distribution? (f) What is the mean age of reproduction in an equilibrium population?

2. Here are two hypothetical populations:

$$
\text{A:}\quad b_0 = 0,\ b_1 = 2,\ b_2 = 3; \quad p_0 = \frac{1}{2},\ p_1 = \frac{2}{3},\ p_2 = 0.
$$

$$
\text{B:}\quad b_0 = 0,\ b_1 = 3,\ b_2 = 4; \quad p_0 = \frac{1}{3},\ p_1 = \frac{3}{4},\ p_2 = 0.
$$

For each population, A and B, (a) What is the equilibrium rate of increase? (b) What is the equilibrium age distribution? (c) What is the total number of offspring per newborn?

3. In a continuous population $1(x) = e^{-cx}$ and $b(x) = k$, where c and k are constants. For numerical examples let $c = 0.02$ and $k = 0.03$. (a) What is the life expectancy of a newborn? (b–f) Answer the questions in Problem 1.

4. In Problem 1 replace b_2 by 0 and b_3 by 3/2. (a) Will this population ultimately increase, decrease, or neither? (b) Will it attain a stable age distribution?

5. Which of the following populations will attain age equilibrium most rapidly? (a) A population that reproduces only at age 20 and 21. (b) A population that reproduces at all ages from 11 to 30.

6. The human world population was 4 billion in 1976, and the average rate of increase was about 2 percent per year. (a) If this rate continues, what will be the population at the turn of the century? (b) How long will it take for the population size to double?

7. A bacterial population grows logistically with parameters r and K. A culture starts with one cell. (a) How long will the population take to reach half its maximum number? (b) Show that at this time the population is increasing at the maximum rate.

8. Human demographic tables often give the mean age of reproduction of a cohort of females born at the same time. Alternatively we could ask for the mean age of reproducing females in a population at age equilibrium. (a) Which category is larger in a growing population? (b) When will they be the same?

9. Some ecologists have theorized that to maximize the probability of a future food supply, prudent predators (for example, wolves feeding on moose) attack prey at ages when the reproductive value of the prey is at a minimum. Does this theory seem reasonable?

10. *Optional:* Verify the first integration at the top of page 166.

11. My favorite equation for the world population, which fits the data for the past century or two quite well, is

$$N = \frac{1.79 \times 10^{11}}{(2026.87 - t)^{.99}}.$$

Any comments?

CHAPTER 7

POPULATION GENETICS AND EVOLUTION

Much of the discussion up to this point has been related to evolution. Mendel plus Darwin plus numerous field and laboratory studies plus mathematical modeling have led to a deep understanding of how evolutionary changes can occur.

In Chapter 1 we saw that Mendelian inheritance is a very effective device for conserving genetic variance and that segregation and recombination provide an effective mechanism for gene scrambling. In Chapter 2 we learned how inbreeding increases homozygosity and how random drift can cause gene frequency changes in small populations. Chapter 3 showed how the geographical structure of a population can influence the amount of within- and between-group variability. Chapter 4 showed some of the enormous variety of ways in which natural selection changes allele frequencies. We saw that in long periods of time even a very mild selective pressure can produce large changes in allele frequencies. We also saw that most selection does not lead to systematic evolutionary changes but rather to maintaining the status quo by removing recurrent mutations and segregants from favored heterozygotes. In addition, we learned that natural selection usually cannot carry a population from one adaptive peak to another.

Chapter 5 extended selection results to polygenic systems. Since most evolutionary changes are quantitative rather than all-or-none, it is important to understand how selection changes such traits. We saw that natural selection acts on the additive, or genic, part of the variance. We saw further that most selection in natural populations is stabilizing, directed against phenotypic outliers. Chapter 6 discussed the complications of realistic life histories. Although this subject is in its infancy, the main conclusion was that populations rather quickly adjust to a stable age distribution, at which time equations developed from simple, discrete-generation models work quite well.

Mutation (including chromosomal changes), Mendelian segregation and recombination, random processes, and natural selection are sufficient mechanisms for intraspecies evolution. In that sense the problem of evolution is solved. Yet when we look beyond this most superficial description, we find many subleties and some unresolved problems. For example, there remains the intriguing question of whether mechanisms that account for short-term changes are sufficient, when extended over very long time periods, to explain the origins of species and higher taxonomic categories or whether new principles are needed.

In this chapter we look at several different evolutionary problems, some old and some new. The subject is enormous, and we can only scratch the surface, as I have done rather unevenly by selecting special topics.

7-1 EVOLUTION AT THE GENOMIC LEVEL: MOLECULAR EVOLUTION

One of the biggest surprises from the study of changes at the level of the individual amino acid or nucleotide has been that the rates of molecular evolution are much less variable than the rates of change of bodily form and function.

Amino Acid Replacement in Hemoglobins. Now that complete sequences of several proteins are known in many species, we are in a position to consider the rates of change. Of course, we cannot determine amino acid sequences for species that have been extinct for tens, or hundreds, of millions of years. Yet by using paleontological data for phylogeny and dates and putting this information together with amino acid differences in contemporary vertebrates, we can infer what the rates have been.

Alpha hemoglobin is a good example. If we examine the sequences of the 141 amino acids in the human and dog, we find that they differ at 23 sites, or 16.3 percent. Additional examples are given in the upper right half Table 7-1. In general, the more

TABLE 7-1

Proportion of amino acids in α-hemoglobin that differ between various pairs of vertebrates. The values above the diagonal line are the proportion of amino acid sites with one or more changes. Those below the line have been corrected for multiple changes and give the mean number of changes per amino acid site.

	SHARK	CARP	NEWT	CHICKEN	ECHIDNA	KANGAROO	DOG	HUMAN
SHARK		0.594	0.614	0.597	0.604	0.554	0.568	0.532
CARP	0.901		0.532	0.514	0.536	0.507	0.479	0.486
NEWT	0.952	0.759		0.447	0.504	0.475	0.461	0.440
CHICKEN	0.909	0.722	0.592		0.340	0.291	0.312	0.248
ECHIDNA	0.926	0.768	0.701	0.416		0.348	0.298	0.262
KANGAROO	0.807	0.707	0.644	0.344	0.428		0.234	0.191
DOG	0.839	0.652	0.618	0.374	0.354	0.267		0.163
HUMAN	0.759	0.666	0.580	0.285	0.304	0.212	0.178	

Data from Kimura. 1983.

species differ in other ways, the larger their amino acid differences. The table ignores a few changes by duplication or deletion and concentrates on substitutions.

Before we consider the actual rates, we need to correct for the possibility that some of the changes are multiple; for example, a change from threonine to serine may really be from threonine to proline to serine. Sometimes we can infer a multiple change from the code if a single nucleotide change cannot possibly transform one amino acid to another. Usually, however, the code does not provide enough information for such a specific adjustment. Elaborate ways of adjusting the data to take multiple changes into account have been devised, but a very simple method of correction works quite well.

Let P be the observed proportion of amino acid sites in which one or more changes have occurred. Assume that the probability of an amino acid change per year is p and that the total time is N years. Assume further that the individual changes are independent. The probability of no change during this time is $(1-p)^N = [(1-p)^{1/p}]^{Np}$. Let $m = Np$, the mean number of changes per site. Note that since p is very small, of the order 10^{-8} or less, the quantity in brackets is very nearly e^{-1}. Thus the probability of no change in N years is e^{-m}. Equating this expression to the observed fraction of amino acids in which no change has occurred, which is $1-P$, gives

$$e^{-m} = 1 - P$$

or

$$m = -\ln(1-P), \qquad 7\text{-}1$$

where P is the proportion of amino acid sites with at least one change. Correcting the data comparing the human and the dog, where $P = 0.163$, we get $m = 0.178$. (You may have recognized that I have based the correction on the Poisson distribution. For a discussion of this distribution, see Appendix 3.)

The numbers in the lower left half of Table 7-1 have been corrected in this way to give the mean number of changes per amino acid site. According to the paleontological record, the ancestry of the dog and human diverged about 100 million years ago. Thus the dog and human have undergone 200 million years of separate evolution, so that the rate of substitutions per year is $0.178/(2 \times 10^8) = 0.89 \times 10^{-9}$—roughly one amino acid substitution per billion years. When we take into account the times since divergence from a common ancestor for the various numbers in the lower part of Table 7-1, all of them turn out to have about one substitution per billion years.

Figure 7-1 brings out this phenomenon in a slightly different way. Here the mean number of differences per amino acid, m, is plotted against the time (twice the time

FIGURE 7-1. The mean number of amino acid substitutions (ordinate) as a function of the number of years of separate hemoglobin evolution. (Adapted from Kimura. 1983.)

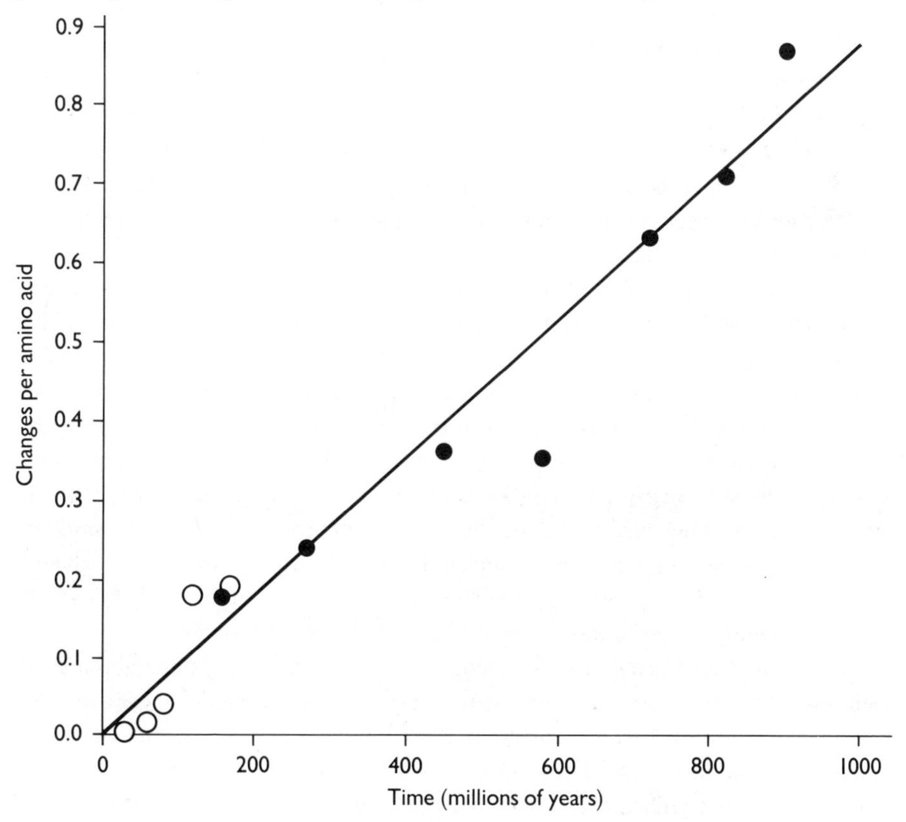

since the most recent common ancestor). The rightmost point is the average of the differences between the shark and the other species (the first column in Table 7-1). The next point is the average of the differences between the carp and the other species (second column); and so on. The slope, b_{YX}, of the regression line is 0.895×10^{-9}, or 0.9 substitutions per amino acid per billion years. The open circles represent data among primates for beta hemoglobin. The average rate is about the same as that for alpha hemoglobin among more distantly related vertebrates.

If the rate of change per amino acid codon is about one per billion years, you might think that the rate of change per nucleotide is about 1/3 as high, since any of the three nucleotides in the codon could change. This simple conclusion is incorrect, however, because many nucleotide changes are synonymous. We shall see later in the chapter that synonymous changes occur at a considerably more rapid rate.

The similarity of rates over widely contrasting species and over greatly different time spans is remarkable in view of the enormous differences in the rates of morphological change. The shark has changed very little during this time, whereas the mammals have changed from a fishlike ancestor to the remarkable diversity of the types we see today.

Another comparison more strikingly brings out this point. If we compare human hemoglobin alpha with beta, 62 amino acids are unchanged and 9 amino acid lengths have duplications or deficiencies (gaps) among the total of 147 positions. Ignoring the gaps, we have 62 unchanged amino acids at 138 positions. If we correct for multiple changes, the mean number of changes per site is $-\ln(62/138) = 0.80$. Comparing carp alpha with human beta gives 61 unchanged positions among 139 sites, plus 10 gaps. The mean number of substitutions per site is then 0.82. The similarity of the two comparisons is evident, both in number of substitutions and number of gaps.

The human globin gene duplicated to form two genes, probably by polyploidization, some 500 million years ago. They have evolved separately since that time. The separation of the ancestry of human from carp is thought to have occurred more than 400 million years ago. That is, the human beta and carp alpha have been in separate organisms for more than 400 million years, subject to the different selection brought about by air versus water environments. In contrast, the human beta and human alpha have been in the same organism — the same environment — for all that time. The two proteins are parts of the same tetramer. Yet the divergence is essentially the same in the two comparisons, despite the enormous differences between humans and fish and the much more rapid morphological change in the human ancestry. Clearly hemoglobin and morphological evolution are marching to different drummers.

The literature contains much discussion on whether or not the rate of amino acid substitution in a specific protein is really constant. Some have argued that each protein has an intrinsic rate, that is, a "molecular clock." Others have noted that when the constancy is examined carefully, it does not stand up to statistical analysis: the variation in rate is significantly higher than we would expect from a pure Poisson process. Yet to me the near constancy is more interesting than the relatively minor departures therefrom and calls for an explanation.

Rates for Other Proteins and for DNA. Each protein seems to have its own characteristic rate of change, and these rates differ greatly. However, the rate is roughly constant for each protein. Table 7-2 gives rates for some different proteins.

The rate is inversely related to the specificity of the individual amino acids for protein function. Histone fits very closely to the DNA molecule, which probably means that each amino acid is important and any substitution causes a serious disruption of function. In contrast, fibrinopeptide, a protein that is discarded when the blood clot is formed, appears to be able to function as well if an amino acid is replaced. Hemoglobin and cytochrome are intermediate in specificity, cytochrome being somewhat more restricted.

The principle that emerges is that those proteins that have the fewest constraints evolve the fastest. This rule can be extended to parts of the molecule. For example, the rate of substitution is about 10 times as fast on the surface of the hemoglobin molecule as in the inner parts close to the heme group. Likewise proinsulin consists of a long polypeptide with three parts — A, C, and B. It later splits into the three parts, of which A and B form the insulin molecule and C is discarded. The evolution rate in the A and B parts is about 0.4×10^{-9} per amino acid per year, whereas in the C part it is 2.4×10^{-9}.

Now that biochemists can determine nucleotide sequences, they have made more discoveries of this kind. Intervening sequences (regions of the gene that are not translated) evolve faster than the translated parts. Pseudogenes (nonfunctional copies

TABLE 7-2

Rates of amino acid replacements in different proteins. The rates are the mean number of replacements per billion years.

PROTEINS	RATE PER BILLION YEARS
Fibrinopeptides	8.3
Ribonuclease	2.1
Lysozyme	2.0
Hemoglobins	1.0
Myoglobin	0.9
Insulin	0.4
Cytochrome c	0.3
Histone H4	0.01

Data from Kimura. 1983.

of genes that have arisen by duplication and have lost their function) evolve faster than functional genes. Finally, because of the redundancy of the genetic code, there are usually several codons for the same amino acid. Synonymous nucleotide substitutions are much more rapid than those that cause the coding of a different amino acid. This phenomenon is especially striking in very slowly evolving proteins such as histone, in which synonymous changes are just as fast as such changes in rapidly evolving proteins. The conclusion is clear:

The smaller the effect that is caused by a nucleotide change, the more rapid that change is in evolutionary time.

The reason for this correlation is surely that those changes that produce a large effect are more likely to be harmful and are eliminated by natural selection.

Evolution by Random Drift. The findings just described show, somewhat paradoxically, that the less strongly a nucleotide or amino acid is selected, the more rapidly evolutionary substitutions occur. This relationship would seem to argue that selection is not the driving force. Yet this conclusion does not necessarily follow, since the smaller the effect of a mutation, the more likely it is to be favorable.

The reason for this correlation is that mutational changes, being random with respect to the functioning of the organism, are usually harmful. They are harmful for the same reason that random tinkering with a watch usually makes it run more poorly or stops it entirely. If the change is slight enough, however, it might happen to make the watch run more accurately, whereas a gross change is almost certain to make it worse. Therefore we are not surprised that the less crucial a nucleotide substitution is, the more likely it is to cause an improvement. Those regions of the molecule that are most crucial can hardly ever be changed without producing something less functional. In this way we can understand, for example, the extreme evolutionary conservatism of those parts of the hemoglobin molecule that are closely associated with the iron-binding heme group.

To evolutionists this result came as no surprise. It is a part of the general Darwinian view of evolutionary gradualism. The argument was developed most convincingly by R. A. Fisher (1958, pages 41 ff). Yet for all its plausibility, this conclusion is difficult to reconcile with all the facts. To discuss this topic, we need to consider the rates of evolution at loci where selection is so weak that random processes must be taken into account.

When we compare the number of amino acid or nucleotide substitutions between different species, we are looking at a process that has taken tens, or hundreds, of millions of years. In Section 2-4 we saw that completely neutral mutants are substituted at a rate equal to their mutation rates. We shall now consider mutants that are nearly, but not quite, neutral.

TABLE 7-3

A model for deriving the probability of fixation of a new mutation. The mutant base or amino acid is indicated by B', or B' can stand for a new mutant allele, whatever its molecular basis is. ($k \geq 1$.)

GENOTYPE	BB	BB'	$B'B'$
Initial frequency	$1 - 1/2N$	$1/2N$	0
Relative fitness	1	$1 + s$	$1 + ks$

As was discussed in that section, the rate of substitution, when viewed over a very long time period, is

$$\text{Rate} = R = 2N\mu P, \qquad 7\text{-}2$$

where, as before, N is the population number, and μ is the mutation rate per gene per generation, and P is the probability that the mutant will eventually be fixed (that is, homozygous throughout the population).

The formula for P was first derived by Haldane, improved by Fisher, further generalized by Malécot, and finally written by Kimura in the form I shall use here. The derivation, which is beyond the scope of this book, is given in C&K, pages 418–430. The model is given in Table 7-3.

Even if the mutant is favorable ($s > 0$), it is still very likely to be lost in the first few generations after it occurs, simply through the random processes of meiosis and the varying numbers of progeny of different individuals. If the mutant is "lucky" enough to increase so that it is represented several times in the population, then it is very likely to become established in the gene pool. Hence the "decision" as to whether the mutant will ultimately be fixed or lost is made while it is still at low frequency, during which time it is represented almost entirely in heterozygotes. Hence the value of k is largely irrelevant, provided it is 1 or larger.

P is given by

$$P = \frac{1 - e^{-2(N_e/N)s}}{1 - e^{-4N_e s}}. \qquad 7\text{-}3$$

Let us consider three cases:

1. Favorable mutant, $s > 0$. If $N_e s$ is much greater than 1, the denominator is essentially 1. Regarding the numerator, recall that $e^{-x} = 1 - x + x^2/2 - \ldots$. Using this expansion and ignoring all terms involving s^2 and higher powers, we obtain

$$P = 2s\,\frac{N_e}{N}. \qquad \text{7-4}$$

(Haldane's widely quoted value of $2s$ was obtained before the distinction between actual and effective number was recognized.) Combining Equations 7-2 and 7-4, we have the rate of substitution

$$R = 4N_e\mu s. \qquad \text{7-5}$$

2. Neutral mutant, $s = 0$. In this case we expand both the numerator and denominator of Equation 7-3, as we previously did for just the numerator, and everything cancels except $1/2N$. Substituting this expression into Equation 7-2 gives

$$R = \mu. \qquad \text{7-6}$$

You may recall that this result is the same as that we obtained by a simple symmetry argument in Section 2-4.

3. Deleterious mutant, $s < 0$. In this case we cannot obtain a simple expression as with the other two cases, but we can easily see that when s is negative, the numerator and denominator are both negative. The absolute value of the denominator increases rapidly as $|s|$ increases. R becomes very small as N_e increases. The results are shown graphically in Figure 7-2.

If the mutant is favorable, we see from Equation 7-5 that the substitution rate is proportional to s, μ, and N_e. The proportionality to s and μ is not surprising. In contrast, we are surprised that for a given value of μ and s, the rate increases with N_e (also brought out in Figure 7-2). It is contrary to fact, for there is no proportionality between molecular evolution rate and population size. Remember that we are concerned with the effective number of the whole species, not that of a local subgroup. For example, insects with enormous population sizes evolve no more rapidly than mammals. Abundant, small mammals do not evolve faster than rare, large ones, nor do common herbivores evolve faster than scarce carnivores. Thus a simple theory based on the probability of fixation of rare, favorable mutants does not fit the molecular data.

Complete neutrality fits the facts better than selection does, but there are two difficulties. One is that we observe the constancy of molecular evolution when we count time in years. Yet mutation rates are thought to be adjusted to the life cycle, so that we should expect constancy per generation rather than per year. Recent evidence showing faster evolution of rodents than of primates (in absolute time) suggests that the rate is related to generation length, thus somewhat strengthening the random drift hypothesis.

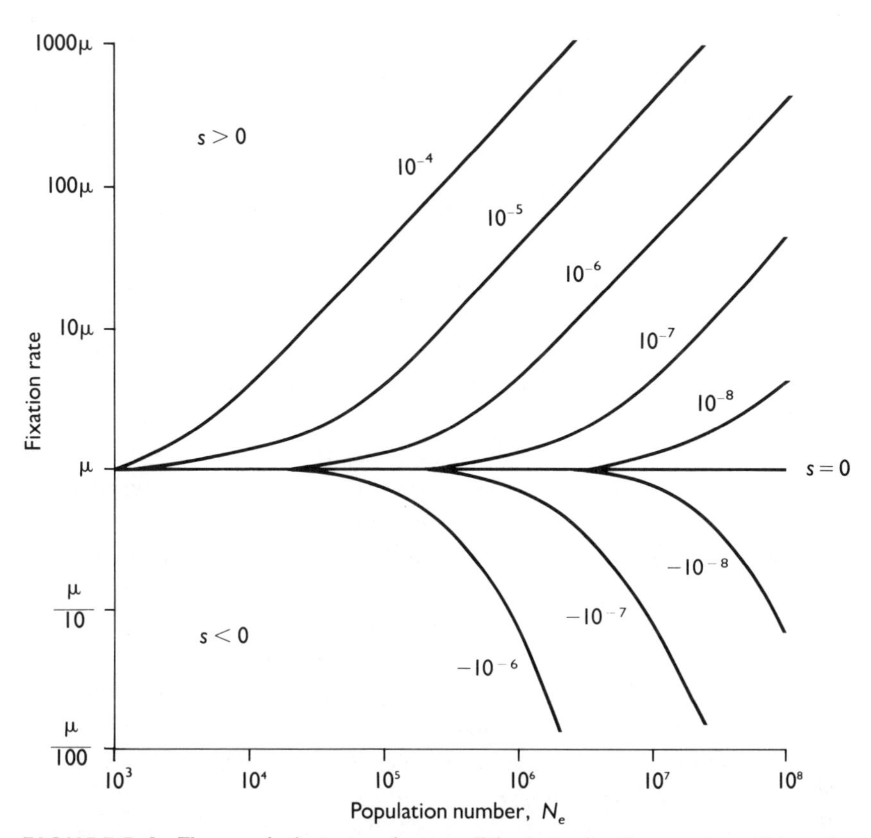

FIGURE 7-2. The rate of substitution of mutant alleles (or nucleotides or amino acids) as a function of the effective population number, N_e, and the selective advantage, s, of the mutant heterozygote. The ordinate is the substitution rate as a multiple of μ, the mutation rate.

Note that the constancy within a lineage may be partly the result of gaps in the record. The constancy may be the average value, which contains appreciable fluctuations that would be observed if the record were more nearly continuous. This variability, however, is consistent with either the selection or drift hypothesis, for we have no reason to think that mutation rates or the proportion of neutral mutants has remained constant. Again the more important point is that the average rate for a given protein is roughly constant from one group of organisms to another.

The second problem is that if mutant substitutions are neutral and we regard molecular diversity as a transient stage in such a process, more polymorphisms should occur in large populations (recall Equation 2-13). Although invertebrates are more heterozygous than vertebrates, they are not as much more as Equation 2-13 would suggest.

The failure of either selectively favored or neutral alleles to fit the data forces us to consider $s < 0$. This is consistent with most of the facts. If the rate per year for a given protein is constant, then the rate per generation increases as the generation length becomes longer. Animals with long generation lengths, the larger ones, tend to have smaller population numbers. Therefore the substitution rate per generation decreases as N increases. This relationship corresponds to the lower part of Figure 7-2, which implies that $s < 0$.

This suggestion was made by Tomoko Ohta and Motoo Kimura (see Kimura, 1983). It has not been very popular because geneticists do not like to think of molecular evolution as a process that steadily lowers fitness. Note, however, that Kimura does not advocate that all evolutionary changes are slightly deleterious and fixed by random drift, only that most of what is observed by molecular methods happens in this way. A very small fraction of favorable mutants could easily offset the decline in fitness.

Fisher (1930, 1958) emphasized that part of natural selection is expended on keeping up with a deteriorating environment. For any species the environment continually gets worse, because of the depletion of resources, overcrowding, and so forth. Even if the species reaches an equilibrium with these deteriorating conditions, the other species do not stay fixed. Competing species are also being improved by natural selection, making competition stiffer for the species we are considering. Again we are reminded of Alice and her treadmill, where "it takes all the running you can do to keep in the same place." From this perspective we could say that the steady deterioration of the genome under mutation pressure is one more component of the treadmill.

I have given only a few of the arguments relating to the theory of evolution by random drift. For a forceful, advocating presentation of the neutral theory, see Kimura (1983).

Most of the alternative arguments invoke selection as the major reason for allelic substitutions. One of the more attractive theories views the continually changing environment as a force causing the fitnesses of different alleles to change with it. A previously deleterious or neutral allele now becomes favorable, and natural selection carries it to fixation. The mutation rate no longer plays a role in determining the rate, unless it is so low that the supply of potentially favorable mutants is insufficient.

This theory is indeed plausible. The main drawback is that it does not suggest any constancy of rates of the same protein in different environments. Can we realistically assume that the rate at which previously deleterious mutants become favorable is the same in fishes as in mammals and birds?

If this discussion seems inconclusive, that is my intention. The neutral theory is inventive and brings together a number of otherwise unrelated observations. It has generated a number of hypotheses and has stimulated a great deal of work, both theoretical and observational. Although acceptance is increasing, it remains controversial. One point is abundantly clear, however: molecular and mathematical studies have shown that any selective forces bringing about molecular changes are at most very, very weak.

Evolution by Gene Duplication. In addition to nucleotide substitutions, along with deletion, duplication, and transposition of one or more nucleotides, several mechanisms increase, decrease, or move larger blocks of DNA. One of the characteristics of animal and plant life is the enormous differences in the amount of DNA in different species. The quantity of DNA is not necessarily correlated with the complexity of the organisms. Some plants have more DNA than any animal, and some amphibians have much more than any mammal. Even closely related species may differ greatly in DNA amount; yet we have no reason to think that they differ much in gene number.

One mechanism for increasing DNA amount, polyploidy, has been recognized since early in the century. Many polyploid plants are known, and often closely related species include diploids, tetraploids, hexaploids, and so forth. Polyploidy is less common in animals than in plants, but many examples are still known, and it is likely that increases in the total genomic DNA by this process have happened several times in the history of animals. For example, the fact that alpha and beta hemoglobin are unlinked is consistent with such an event, followed by divergence of the two types.

Individual chromosomes can also become duplicated by nondisjunction or other accidents of chromosome behavior. Usually trisomic types are deleterious and are quickly eliminated by natural selection. But exceptions do occur, and in some instances chromosomes and DNA have increased in this way, particularly chromosomes that seem to have very few, if any, essential genes. Many plants have B chromosomes, usually with no apparent function, but which are carried along during cell division, increasing the total amount of DNA.

At a finer level, errors in crossing over can produce duplication of chromosome material. Such errors can also produce complementary deletions, but these deletions tend to be deleterious and quickly eliminated. There are many examples of closely linked clusters of identical or very similar stretches of DNA. Errors in crossing over are a sufficient explanation. As with the *Bar* duplication in Drosophila, once a duplication occurs, out-of-register synapsis is frequent, so that triplication and further changes become much more likely.

Other mechanisms, which are not too well understood as yet, allow the DNA to increase itself by a type of local copying. This process can lead to many duplications of a particular region. The human genome contains many so-called *aleu* regions. These regions are homogeneously staining blocks of repeated DNA units. Their function, if any, is not known; they may simply be consequences of DNA behavior.

Transposable elements are stretches of DNA, a few kilobases to tens of kilobases in length, that have the capacity to move from place to place in the genome. They usually include genes coding for the enzymes required for transposition. Often there are mechanisms, either internal to the element or elsewhere in the genome, that regulate the transposition rate. If such an element moves to a functional gene, it usually disrupts that function and is detected as a mutation would be. Precise excision of the transposed region restores the original function. In higher organisms the element usually moves to a region with no function and becomes a part of the chromosome. Later this element may lose its capacity to move and simply makes the chromosome longer than before. What makes transposable elements particularly significant from this standpoint is that many have the capacity to move to a new site while leaving a copy of themselves at the old site. Thus there is a built-in mechanism for increasing the amount of DNA in the cell.

Recent evidence implicates retrovirus DNA and reverse transcriptase as a mechanism for transposition. Most of the pseudogenes studied so far lack the intervening sequences of the parent gene, which strongly supports the argument that the transposition somehow involved processed RNA—hence the implication of retroviruses and reverse transcriptase in the process. For example, the human gene for agininosuccinase has more than a dozen pseudogene copies, all lacking intervening sequences and scattered on several different chromosomes.

A persistent puzzle brought out by recent advances in cytological and molecular techniques is the large amount of DNA that has no known function. The amount can differ enormously from species to species, even among closely related ones. We have every reason to think that there is no substantial difference in gene number among these species; the difference is in "junk" DNA. Figure 7-3 shows a particularly striking example.

The fourth chromosome in *Drosophila melanogaster* is much smaller than any of the other chromosomes. In another species, *D. ananassae,* it is considerably larger,

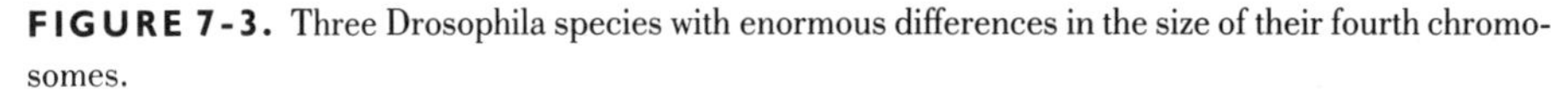

FIGURE 7-3. Three Drosophila species with enormous differences in the size of their fourth chromosomes.

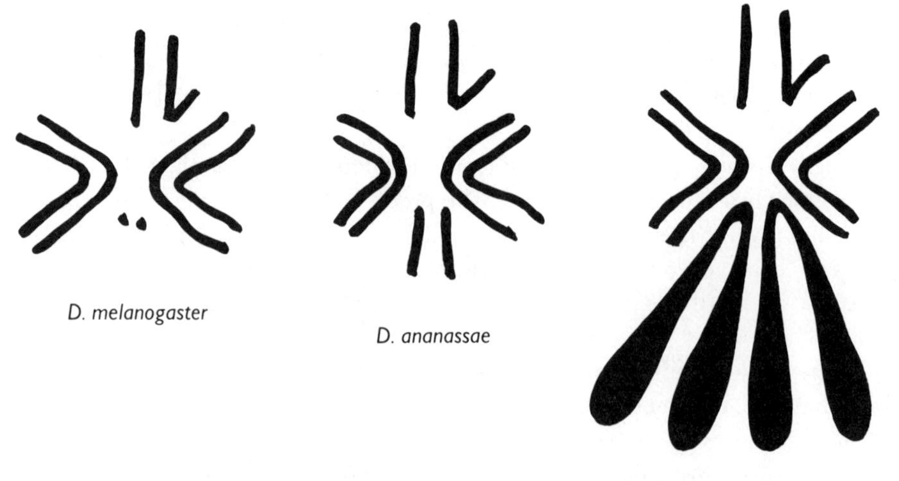

although still relatively small. In contrast, the fourth chromosome of *D. nasutoides* is enormous, much larger than any of the other chromosomes. The extra material is all homogeneously staining and made up of multiple copies of the same basic unit. This is the most extreme example of runaway DNA duplication that I know of.

The study of mechanisms of transposition and multiplication of DNA is currently being very actively pursued. Research is still at too early a stage to provide detailed data; in addition, such a discussion would be out of place in this book. Our emphasis here is on understanding the evolutionary consequences of such processes rather than on the processes themselves.

Clearly the amount of DNA can increase in many ways. It is hard to imagine that this increase benefits the organism; rather it increases the amount of DNA that must be copied, consumes energy, and complicates mitosis and meiosis. Most likely the explanation lies not in the organism but in the DNA itself. If bits of DNA and RNA have the intrinsic capacity to make copies of themselves and to move these copies around the genome, the organism can do nothing about it. If the impairment is too great, the organism dies.

Thus DNA and RNA are analogous to a parasite. We would expect, then, that the DNA will go its own way, increasing in amounts according to whatever type of **parasitic,** or **selfish, DNA** it possesses. This viewing of selfish DNA from the standpoint of the DNA itself pushes natural selection to a lower level of organization than we usually think of, but the principles should be mostly the same. The most successful parasitic DNA, like successful parasites in general, is the one that best perpetuates itself while not greatly reducing the fitness of its host, which would cause its own extinction.

Several recent articles have attempted to develop mathematical theories for selfish DNA. Thus far the mechanisms are so varied that simple, widely applicable principles are hard to detect. I shall not discuss such work in this book, although this subject will very likely become an important part of population genetics theory in the future. The theory presented in this book is based on very solid facts — Mendelian inheritance and the linear order of genes on the chromosome. It seems premature to include a mathematical theory of transposition, selfish DNA, and multi-gene families in this book while the underlying mechanisms are still being worked out.

Meiotic Drive. Another mechanism affecting the composition of the genome by means other than organismic natural selection is based on systematic deviations from Mendelian ratios. Instead of transmitting an equal number of the two kinds of genes, the *Aa* heterozygote transmits an excess of one. This process is called **meiotic drive.** Genes perpetuating themselves by meiotic drive may be colorfully described as cheating genes or as genes that violate Mendel's laws.

The mechanisms are varied, and not all of them involve meiosis. Some chromosomes are transmitted in excess of their Mendelian expectation by being preferentially included in the functional egg instead of the polar body. In Drosophila females, if a pair of homologs differ in length, the shorter one is more often included in the nucleus that gets fertilized. Although these mechanisms cause changes in gene frequencies, they are not thought to be of much evolutionary significance.

B chromosomes in maize are small, heterochromatic chromosomes with no known function. They tend to increase in number up to the point at which they begin to have a deleterious effect on plant viability. There is often nondisjunction of B chromosomes during the final cell division in the pollen tube. Somehow the nucleus with the extra B chromosome regularly fertilizes the egg, while the other nucleus goes into the endosperm. In this way the B chromosomes increase in number until an equilibrium between the transmission bias and the deleterious effect is eventually reached. We can write equations quite similar to those for mutation-selection balance, and the dynamics are much the same.

One example of meiotic drive occurs at the *T* locus of the house mouse. Mutants in this region (several closely linked genes are involved) affect the tail and also cause a variety of embryological abnormalities. Despite these deleterious effects, *T* alleles are often found in natural populations. They are kept at rather high frequencies by the fact that heterozygous males produce a large excess of functional sperm carrying the *T* allele.

The most thoroughly studied example of meiotic drive is the *SD*, for **segregation distorter,** region in *Drosophila melanogaster*. The *Sd* allele causes no visible effect, yet it is transmitted to 95 percent or more of the progeny. The mechanism is through interference in the normal process of sperm maturation. Somehow the *Sd* allele "instructs" the homologous chromosome to self-destruct. *Sd* acts through a locus on the homologous chromosome, R^s, for "responder sensitive." The killer *Sd* chromosome can be converted into a suicide chromosome by being in cis phase with a sensitive responder, R^s.

The evolutionary interplay of these loci makes a fascinating story. The chromosome in the original population is thought to be $Sd^+ R^s$. Somehow a double mutant, $Sd\ R^i$ (a segregation distorter linked to an insensitive responder), arose. This chromosome would spread through the population very rapidly. By rare recombination an $Sd^+ R^i$ was produced, and it is immune to distortion. It will therefore increase in the population. The distorting chromosome is somewhat harmful in other respects, so that as long as its distorting activity is neutralized by insensitive chromosomes, it tends to decrease. But when it is rare, the insensitive chromosome, also disadvantageous by itself, decreases. Thus we have the kind of cycle shown in Figure 7-4.

Is meiotic drive an important evolutionary phenomenon? If a chromosome causing some form of segregation distortion has no harmful effects otherwise, we would expect it to sweep through the population to fixation, carrying along whatever hitchhiking genes might be linked to it. This situation is probably quite rare, although we do not know for sure.

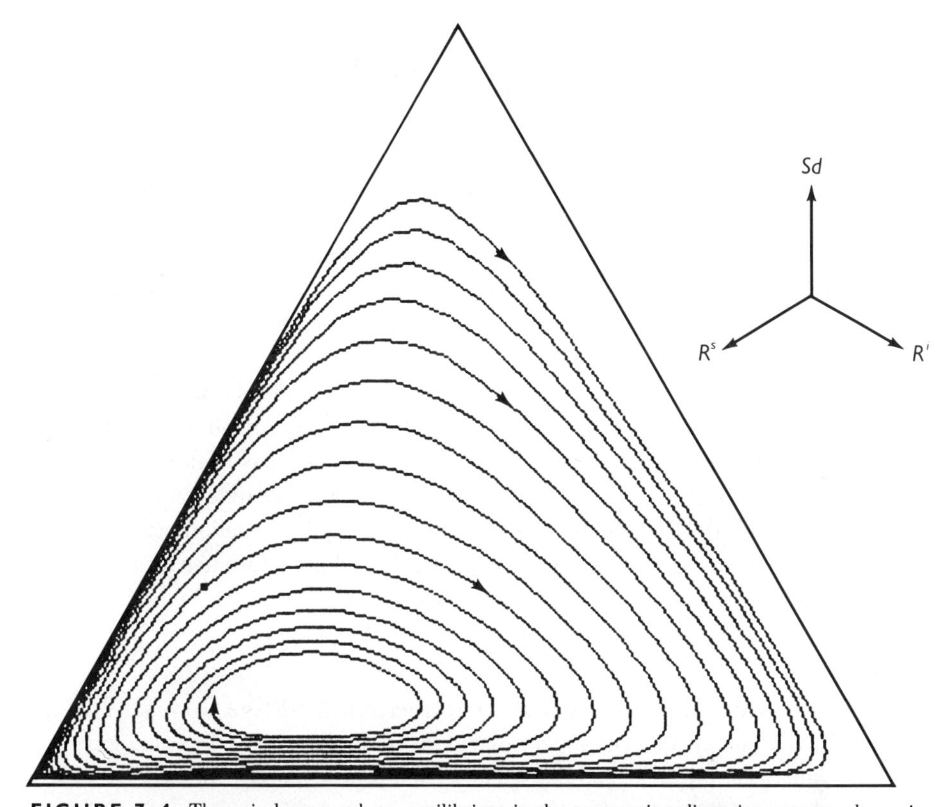

FIGURE 7-4. The spiral approach to equilibrium in the segregation distortion system, shown in homogeneous triangular coordinates. The perpendicular distance from the bottom side of the triangle is the frequency of *Sd* R^i, from the left is that of Sd^+ R^i, and from the right side is that of Sd^+ R^s. The population started at the lower left, with mainly the normal chromosome (Sd^+ R^s), and spirals inward toward an equilibrium.

Because of the ubiquity of pleiotropy, the drive mechanism probably has other effects that are harmful. If so, the increase of the driven chromosome has its price. Perhaps species are harmed this way, even driven to extinction.

Equilibria may occur similar to those of the *SD* system. Although the driven chromosome is kept at fairly low frequency in the population, it nevertheless causes some harm. Yet the drive mechanism prevents the population from getting rid of it. It could be a mechanism for keeping some closely linked harmful genes in the population at higher frequencies than mutation-selection balance would predict. Perhaps, although I think not likely, some serious human genetic diseases (cystic fibrosis, for example) that are too common to explain by conventional mutation rates are maintained by meiotic drive. The alternative explanation, which I suspect is more likely, is that some form of selection favors the heterozygote, as was discussed in Chapter 4.

7-2 EVOLUTION OF FORM AND FUNCTION

The principles of selection on genetic traits, either qualitative or quantitative, are the same whether selection is natural or artificial. If nature selects for large size, that is, if larger individuals have a greater probability of surviving and reproducing, then size will increase in proportion to the intensity of the selection and the narrow sense heritability of the trait. Selection under human control can of course concentrate on specific traits, such as rate of gain in swine, and can enhance the information about the genotype from observation of the performance of relatives. In nature the only criterion for selection is Darwinian fitness.

As emphasized before, most selection for qualitative traits is static—removal of deleterious mutants, maintenance of polymorphisms, and adjustment to transient environmental differences. Most selection for quantitative traits is stabilizing. Mean fitness, $\bar{w}$, is expected to be very nearly one, since most of the traits contributing to it are likely to be fairly close to equilibria and the average population size is usually nearly constant. Yet there is likely to be considerable additive variance for almost any measurement or function; a change in the environment that shifts the optimum can lead to rapid change, at least rapid in the context of evolutionary times.

In Praise of Multiple, Independently Inherited, Additive Genes. Although at present we cannot determine the extent of neutral changes in evolution, it is becoming increasingly clear that the selective differences among "normal" alleles at most loci are very slight. This fact leads to the following set of conclusions, which have been long accepted by many evolutionists:

1. The genes on which normal variability depends cause small effects and are very weakly selected. (Of course, this conclusion does not say that major mutations do not occur; they obviously do. But it does say that in the main such genes do not contribute to normal variability and are poor candidates for causing evolutionarily important variance.)
2. The smaller the effect of an allele, the more nearly additive it is likely to be, both within and between loci. This conclusion is expected on theoretical grounds and is observed empirically. The best data come from genes affecting viability in Drosophila.
3. The more nearly additive the genes influencing a trait are, the more responsive the trait is to selection. Recall that selection acts on the additive component of the genetic variance, not the genotypic variance.
4. The more numerous the genes determining a trait are, the less random fluctuation there is from Mendelian segregation and recombination. The

Mendelian "noise" from individual loci largely cancels out, and the process becomes less probabilistic and more nearly deterministic.

5. The more numerous the genes contributing to a trait are, the greater are the possibilities for fine tuning. If the optimum size of the eyeball is 1.75 cm and the available genotypes cause sizes 1.5, 1.6, 1.7, 1.8, and so forth, the optimum can never be achieved. With a larger number of genes, the right combination is more likely to exist.

6. Many segregating genes with roughly additive and more-or-less interchangeable functions offer the possibility for having enormous potential variability with small overt variability. Such a population can respond rapidly to a shift of the optimum and, if the environmental change is continuous and large, produce phenotypes far exceeding the existing range, as the corn and mouse selection experiments described in Chapter 4 show.

Of course, nature is not so simple. Dominance and epistasis frequently intrude and complicate selection. If the chromosome number is small, significant linkages may occur. Pleiotropy is the rule. Characters are correlated; selection for one trait may cause maladaptive changes in another because of pleiotropic or allometric relations. Embryological study brings out many examples of developmental interdependencies. In view of all these complications and interactions, it is surprising that there is as much additivity and independence as is observed.

As I mentioned earlier, for natural selection to work, gene action must have a certain amount of predictability. A gene that causes an increase in size in one background genotype usually does so in others; otherwise selection experiments would produce changes in a direction opposite to the direction of selection or, at best, no change at all. Also, different effects must be sufficiently independent, so that one character can be selected without too many unwanted changes to others. Despite the ubiquity of pleiotropy, it is generally true that we can select for blood pressure without producing unwanted changes in the liver or brain.

Evolution is a composite of many different types of changes. Those alleles that are most important for survival and fertility will increase, and the traits that they determine will increase. Some other traits may be carried along, but I suspect that almost any pleiotypic coupling can be broken. Modifying genes are everywhere and can alter part of a pleiotropic complex. If there are genes with two or more effects, there are usually other genes with one effect or the other, or modifiers that can accomplish the separation.

The Amount of Variance Required for a Gene Substitution. From molecular studies we have a rough idea of how fast genes are changing. I have already emphasized that many of these changes may be the result of random processes, but for the moment let us consider only selected alleles. How much genetic variance is required to accomplish the replacement of one allele by another?

We can determine this amount by resorting to the fundamental theorem of natural selection. From Equation 4-17, assuming that selection is weak so that $w \approx 1$,

$$\Delta\overline{w} \approx V_g. \tag{7-7}$$

Because of the substitution of this allele, the population fitness changes from its initial value $\overline{w}_0$ to $\overline{w}_t$ (or in the terminology of Chapter 4, a change of amount s), where t is the number of generations required for the change to occur. This value is the same as the change per generation summed over the whole time. Thus

$$\overline{w}_t - \overline{w}_0 = s = \sum \Delta\overline{w} = \sum V_g. \tag{7-8}$$

We have a remarkable result:

The total variance in fitness required to accomplish an allele substitution is equal to the fitness change brought about by that substitution.

Since variances are additive over loci (assuming independence, which is a reasonable approximation for multifactorial traits), we can ask for the total of several substitutions at different loci. As an example suppose that 10^4 loci are evolving at a rate of 2.5×10^{-6} gene substitutions per locus per generation, corresponding roughly to the rate for hemoglobins, and each substitution confers an increased fitness of 0.01. The total variance required for each substitution is also 0.01, by Equation 7-8. The average variance required each generation to evolve at this rate would be $2.5 \times 10^{-6} \times 10^{-2} \times 10^4 = 2.5 \times 10^{-4}$. (We have assumed that these 10,000 loci are evolving simultaneously and that the total variance is distributed evenly to each generation.) This variance of fitness is very small. Drosophila data suggest that the additive variance for viability alone is about 0.01, and for total fitness is several times as high.

The conclusion of this discussion is that a very small amount of additive genetic variance is sufficient to account for observed rates of allele substitution, even if all gene substitutions are selective. This conclusion reinforces the view that most of the variance is used to select for maintaining the status quo rather than bringing about evolutionary changes.

Rates of Morphological Evolution. In contrast to molecular changes in specific proteins or DNA as a whole, evolution of form and function occurs at vastly different rates. We and the sharks evolved from a common ancestor some 400 to 500 million years ago. This ancestor was very similar to a shark—at least very much more like a shark than a human. During the time that some sharks have hardly changed at all, the birds and mammals have emerged and have changed in all the ways we see. To take another example, some insects have hardly changed at all in a still longer time period. Thus a first contrast is that average rates in different phylogenies differ enormously.

A second contrast is in the variability of rates within a phylogeny. This point was first emphasized by G. G. Simpson in his book *Tempo and Mode in Evolution* (see Simpson, 1953). Recently the view has been reemphasized and exaggerated by Gould and Eldredge (1977) under the name "punctuated equilibrium." They assert that evolution is characterized by long periods of no observable change, interspersed with sudden jumps. Just how jerky the transitions are is a matter of considerable debate, but the rates are undoubtedly vastly unequal at different times in the same phylogeny.

The Opportunism of Evolution. What circumstances lead to very rapid evolution? One is human intervention. Changes in domestic animals are spectacular both in their extent and in their rapidity. Darwin was puzzled by the extreme variability in dogs and other domestic animals. Now we know that the ancestors of dogs were not particularly variable; the variability that led to such bizarre results lies in human tastes, not canine genes. A change in the direction of selection can lead to rapid change. Other examples, also related to human activity, are industrial melanism and insecticide resistance, discussed in Chapter 4.

In natural habitats some of the most spectacular examples of variability are found on islands. The birds on both the Galápagos and Hawaiian Islands have been studied extensively. In both cases the birds are all finches but have very striking morphological differences, which are especially noticeable in the bills. Some are large for fruit eating, some strong for seed crushing, some long for insect catching, some curved for nectar collection, some strong for wood pecking, and so on. Yet all of these types are derived from a finchlike ancestor not very long ago by geological standards. These birds somehow migrated to islands previously uninhabited by birds — perhaps they were blown in by a storm. All these empty niches were available, and the birds quickly evolved to fill them. On the mainland the niches were already filled; a remodeled finch cannot compete very well with long-adapted woodpeckers.

Equally spectacular are the Hawaiian Drosophila species. For one thing the number of species is enormous; the conditions on these islands must somehow aid in the group isolation that favors a high rate of speciation. The Drosophila are adapted to many kinds of niches, with a corresponding wide variety of morphological modifications. There are even behavioral changes; some species show a distinct territorial behavior, with males butting each other almost like rams.

G. G. Simpson was the first to emphasize this opportunistic feature of evolution, although it had been implicit in the writings of many, including Darwin. Evolution takes advantage of whatever is offered by way of an opportunity to exploit a previously unexploited ecological situation.

In summary, morphological evolution (in striking contrast to molecular evolution) is strongly dependent on environmental opportunities, which determine both the rate and direction of evolutionary change.

Cytological Evolution. A reciprocal translocation follows the same rules for unstable equilibria as were discussed for maternal-fetal incompatibility in Chapter 4. If one of the chromosome pairs is in the original order whereas the other pair is translocated, several unbalanced types of gametes can be produced, leading to inviable offspring. Therefore translocations are ordinarily strongly selected against. However, if a translocation becomes homozygous, the chromosomes pair normally and no unbalanced gametes are produced. The difficulty in going from homozygous normal to a homozygous translocation is that the unstable equilibrium prevents it. Whichever type is more common is favored. The situation is similar with inversions. Crossing over within a heterozygous inversion leads to dicentric and unbalanced gametes; so inversions are also selected against unless they are very common, in which case they are selected for.

Nevertheless translocations and inversions are common features of evolution, especially in the evolution of mammals. Most of them are whole-arm translocations or centromeric fusions and fissions — all types that lead to a minimum of aneuploid gametes. Yet they are sufficiently disadvantageous that it is hard to understand how they evolve by natural selection.

The only likely mechanism is that translocations in very small populations, where selection is ineffective, sometimes drift into high frequencies. Such an event does not happen often, but occasionally it might. The most convincing scenario is that a population goes through a size bottleneck and that a translocation, which just happens to be there at the right time, drifts to a high frequency. Then if the translocation is above the unstable equilibrium, it is maintained by selection when the population becomes large again.

The point of greatest evolutionary interest here is that translocations and inversions are much more frequent in mammalian evolution, particularly in primate evolution, than in that of lower vertebrates and invertebrates. It is the mammals, especially the primates, that have evolved most rapidly.

This fact argues strongly that population bottlenecks and random processes have been important influences in those species in which morphological change has been the fastest. It offers some support to the evolutionary theory of Sewall Wright, to which I now turn.

7-3
THE SHIFTING-BALANCE THEORY OF WRIGHT

Since 1931, Sewall Wright has been concerned with the difficulty of moving from one harmonious combination of alleles to another without passing through less well-adapted types. Suppose, as in Section 4-10, that genotype *aa bb* is fit and that genotype *A– B–* is better, but the two types *A– bb* and *aa B–* are inferior to either. Suppose that *aa bb* is the prevailing type. Unless the alleles happen to be closely linked, a rare *Aa Bb* genotype cannot increase. Its mates will usually be *aa bb*, and the less fit genotypes will be produced by recombination. The favored genotype has to pass over an unstable equilibrium, analogous to that for underdominance at a single locus.

If there are variable dominance, linkage, and multiple loci, the fitness surface (see Figure 4-9) is much more complex and multidimensional. We can visualize a fitness surface with numerous peaks and valleys. A population hung up on a low peak cannot leave it to get to a higher one. A peak need not be at a corner; the optimum genotype may include heterozygotes at one or more loci.

Figure 4-9 grossly exaggerates the vertical dimension. Ordinarily the kinds of genes that are important in evolution are those that have very small effects. Therefore the surface is much flatter, and the peaks and valleys less sharp. If the effects are small, for example, components of several quantitative characters, considerable random drift of allele frequencies will occur.

Wright argues that the best way to avoid being hung up on a low peak is to have the population broken up into many nearly isolated subpopulations. As we saw in Chapter 3, weakly selected alleles will tend to drift in different directions. Wright's theory has two phases. In the first phase the allele frequencies drift to some extent in each of the subgroups. One subgroup might by chance drift into a set of gene frequencies that correspond to a higher peak. Then the second phase sets in. This subgroup now has a higher average fitness than other subgroups and will tend to displace them by an increase in group size and a one-way migration into other groups, until eventually the whole population has the new, favorable gene combination. Then the whole process starts again. In Wright's view such a process permits the evolution of novel types of gene interactions that could never happen by mass selection alone.

Fisher, in contrast, argued that no such theory is needed. In a highly multidimensional fitness surface, the peaks are not very high, the slopes are not steep, and the peaks are connected by fairly high ridges, always shifting because of environmental changes. According to Fisher, the analogy is closer to waves and troughs in an ocean than to a static landscape. Alleles are selected because of their average effects, and a population is unlikely to ever be in such a situation that it can never be improved by direct selection based on additive variance.

One objection to Wright's theory is that it requires a rather special population structure. Furthermore during the second phase, when the favored subgroup is expanding and sending migrants to other groups, recombination will occur with individuals from the other subgroups, which will break down the favorable combinations that arose by chance when the subgroup was isolated. Therefore the requisite conditions for Wright's theory may not often be found. On the other hand, the correlation between morphological and cytological evolution argues that random processes are associated with large morphological change.

I should emphasize that the difference between these two views is not mathematical, although much of Wright's writing is mathematical. The essential difference is physiological: Does going from one favored combination of alleles to another often necessitate passing through genotypes that are of lower fitness? Wright believes that this course is typical and that successful organisms have had to find a way to get around this problem. Fisher argued that evolution has typically proceeded by a succession of small steps, each of which is selectively advantageous, leading eventually to large differences by the accumulation of small ones. According to this view, the best population is a large, panmictic one in which statistical fluctuations are slight and each allele can be fairly tested in combination with many others alleles. According to Wright's view, a more favorable structure is a large population broken up into subgroups, with migration sufficiently restricted (somewhat less than one migrant per generation) and the size sufficiently small to permit appreciable local differentiation.

Fisher and Wright may both be correct: most selection that makes minor modifications to improve fitness follows Fisher's process, whereas occasional new novel types involve a random element and occur by a Wrightian process. A half-century after Wright's theory was put forth, the issue is still not settled.

7-4 SOME EVOLUTIONARY PROBLEMS OF SPECIAL INTEREST

Most of the changes in nature are easily accounted for by mutation, natural selection, and random processes, perhaps aided by a favorable population structure. Yet in many circumstances we must look at the problem differently. For example, natural selection does not always increase individual fitness. Let us look at a few situations of special interest.

Selection for the Sex Ratio. We can easily explain much of natural selection. The strongest, hardiest, most fertile individuals survive and transmit genes for the traits that give them these qualities. Darwin realized this, of course. At the same time he was unable to answer one problem, which he said he would leave for future generations. This problem is how selection leads regularly to a 1 : 1 sex ratio. Such a ratio is especially surprising in highly polygynous species, in which producing many more females than males would seem to make sense. Of course, we now know about the X-Y chromosome mechanism, but many experiments have shown that this mechanism can be easily modified, and species with other mechanisms also have a 1 : 1 ratio.

The problem was solved by Fisher, but I am going to arrive at his conclusion in a different way that I think is easier to understand. Suppose that the population now has more males than females. In this population a female will produce more offspring, on the average, than a male, for the simple reason that the total number of successful eggs and sperms must be equal. Now suppose that in this population a genotype causes parents of either sex to produce an excess of female offspring. The individuals with this genotype will not produce any more progeny than the others, but they will produce more of that sex that produces more progeny. Hence they will have more grandprogeny, and the genes causing the tendency to produce more females will increase. To the extent that the trait is heritable, the sex ratio will shift in the direction of more females. This shift will continue until the population arrives at a 1 : 1 ratio. The same argument works in reverse for a population with an excess of females: alleles tending to augment the proportion of males will increase. Thus there is a stable equilibrium at a 1 : 1 ratio.

This verbal argument is convincing, but it is not specific enough to answer the question as to the stage in the life cycle in which the 1 : 1 ratio is achieved. Is it newborns or individuals of reproductive age? A more concrete way to look at the problem is this:

Suppose that in generation 0 a population of adults produces a total of N progeny, of which $N\bar{p}$ are males and $N\bar{q}$ are females. Among the progeny a subset of phenotype *SR* (for sex ratio) produces a proportion p of males and q of females, and this group produces n progeny. To restrict the problem to selection acting solely through the sex ratio, assume that this genotype has the same viability and fertility as the population in general. Let v_m and v_f be the viability (survival to reproductive age) of males and females, and f_m and f_f the fertilities. Note, however, that the f's will not be constants, since the fertility of the two sexes is determined by the sex ratio of reproducing adults. The situation is set forth in Table 7-4.

From the table the ratio of generation 2 zygotes descended from *SR* genotypes to the numbers in generation 1, which I shall designate as w_{SR}, is

$$w_{SR} = \frac{n(pv_m f_m + qv_f f_f)}{n} = pv_m f_m + qv_f f_f \qquad \text{7-9}$$

and for the population as a whole is

$$\bar{w} = \bar{p}v_m f_m + \bar{q}v_f f_f. \qquad \text{7-10}$$

TABLE 7-4

Effect of selection on the sex ratio.

		PROGENY OF *SR*'s		ENTIRE POPULATION	
GENERATION	STAGE	MALES	FEMALES	MALES	FEMALES
1	Zygote	np	nq	$N\bar{p}$	$N\bar{q}$
	Adult	npv_m	nqv_f	$N\bar{p}v_m$	$N\bar{q}v_f$
	Gametes	$2npv_m f_m$	$2nqv_f f_f$	$2N\bar{p}v_m f_m$	$2N\bar{q}v_f f_f$
2	Zygotes	$n(pv_m f_m + qv_f f_f)$		$N(\bar{p}v_m f_m + \bar{q}v_f f_f)$	

The average excess of fitness for the *SR* genotype (comparable to Equation 4-9) is

$$E_{SR} = \frac{w_{SR} - \bar{w}}{\bar{w}} = \frac{pv_m f_m + qv_f f_f - \bar{p}v_m f_m - \bar{q}v_f f_f}{\bar{p}v_m f_m + \bar{q}v_f f_f}. \quad 7\text{-}11$$

But since the total contribution of males must equal that of females,

$$\bar{p}v_m f_m = \bar{q}v_f f_f. \quad 7\text{-}12$$

Utilizing Equation 7-12 to simplify Equation 7-11 we obtain

$$E_{SR} = \frac{p}{2\bar{p}} + \frac{q}{2\bar{q}} - 1 = \frac{(p - \bar{p})(\bar{q} - \bar{p})}{2\bar{p}\bar{q}}. \quad 7\text{-}13$$

We need not be specific about the inheritance of *SR* or of the mating system. Assuming only that the *SR* genotype has a narrow-sense heritability greater than 0, selection will change the causative genes in the direction of the average excess.

There are two ways for E_{SR} to be 0. One is when $p = \bar{p}$. This condition simply says that if the *SR* genotype is not different from the rest of the population, no change will occur. The other condition tells us something interesting: the *SR* genotype makes no differential contribution when $\bar{q} = \bar{p}$. When the male and female zygotes are equally frequent, the population is at equilibrium.

We can easily see that the equilibrium is stable by examining the sign of E_{SR}. When the population has an excess of females ($\bar{q} > \bar{p}$), then E_{SR} is positive when $p > \bar{p}$. In words, when there is an excess of females, a genotype that produces more males than the population average is favored, and vice versa. To the extent that the trait is heritable, the sex ratio moves toward equality. Thus there is a stable equilibrium when the population sex ratio is 1 : 1.

Notice that $p:q$ is the sex ratio at the zygote stage. Thus this mathematical excursion has shown us what the verbal argument did not—that the sex ratio equilibrates with equality at the zygote stage. Notice that sex differences in viability do not affect this argument; the v's can take any values.

However, pre-adult viability differences in the two sexes can make a difference in some situations. Suppose that, as in humans, males are more likely to die as embryos or infants than females are. As a result a female who produces an excess of sons will, because of their greater death rate, produce her next child sooner than if she produced more daughters. Thus a female that produces an excess of males will produce more zygotes, although fewer adult progeny, than other females. We can show that this situation causes the population sex ratio to equalize at a somewhat later stage, which is the age when the death of a child no longer hastens the conception of the next child.

This argument applies only to species in which a juvenile death hastens the birth of the next offspring. It would not, for example, apply to litter-bearing animals.

This analysis can be made by an extension of the equations just given (see C&K, pages 291–293). The original argument of Fisher, which I, for one, find hard to understand, is given in his book *The Genetical Theory of Natural Selection* (1930; 1958, pages 158 ff).

An interesting aspect of the sex ratio argument is that it shows that natural selection does not maximize population fitness. If it did, an excess of females would be produced by polygynous species. Another example in which natural selection leads to lower fitness is the development of self-fertilizing plants (see Chapter 2). It is important to realize that natural selection is not necessarily a fitness-optimizing process, although it often is.

Sexual Selection. Many traits differ strikingly in the two sexes, in ways that seem to be unrelated to survival. Some examples are the extreme size differences in male and female walruses, the mane of a lion, and beards in one sex of humans. Still more striking are traits that appear to be highly disadvantageous from the standpoint of survival; an example is the tail of a peacock, which must advertise the bird's location to its predators.

Darwin was the first to point out, as he was for so many other ideas, that such traits are the result of what he called **sexual selection,** that is, competition for mates. Deer antlers may be used for fighting other males. Tail feathers and mating displays in birds attract the opposite sex. To leave more progeny than others of their sex, animals have developed extremely varied, and often highly inventive, behavior. For example, the male of some insect species is able to remove sperm deposited in a female by a previous mate before introducing his own sperm.

Sexual selection is often contrary to what would optimize the survival of either the individual or the species. It provides another example of the fact that natural selection does not optimize fitness as we ordinarily think of it.

Cooperative Behavior. Social behavior is often cooperative or altruistic. Cooperative behavior is easy to understand from the standpoint of natural selection: I help you, you help me, and we both profit by it. The problem is more complicated if the cooperative behavior is profitable only when many individuals engage in it. An example is musk oxen, who fend off attacks from wolves by arranging themselves in a circle with their heads on the outside. It is hard to imagine how the behavior pattern first emerged. The first individual that behaved this way instead of running away would not have had much chance for survival. (I don't intend to imply that I know anything about how this behavior pattern actually arose; it is only a hypothetical example.)

We considered frequency-dependent fitnesses in Section 4-8. Here is another example. Assume that the cooperative behavior confers an intrinsic disadvantage, s,

TABLE 7-5

A simple model for a trait that is disadvantageous when rare but advantageous as it becomes common. ($0 < s < 1$; $b > 0$.)

GENOTYPE	AA	Aa	aa
Frequency	p^2	$2pq$	q^2
Relative fitness	1	1	$(1 - s)(1 + bq^2)$

but also an advantage that increases in proportion to the number of individuals with this behavior. Although being cooperative has a cost, it brings a reciprocal benefit when the trait becomes common enough to provide several cooperators. The increased benefit is measured by b. To be explicit, assume that the trait is caused by a recessive allele. This example is set forth in Table 7-5.

Using the methods of Chapter 4, we can write the frequency of the A allele in the next generation:

$$p' = \frac{p}{p^2 + 2pq + q^2(1 - s)(1 + bq^2)}. \tag{7-14}$$

At equilibrium, $p' = p$. Solving the resulting equation leads to

$$\hat{q}^2 = \frac{s}{b(1 - s)}. \tag{7-15}$$

As in some of the examples in Chapter 4, such an equilibrium is unstable. Once the gene for cooperation becomes common enough to exceed the equilibrium point, it will continue to increase, but it cannot increase by natural selection when it is rare.

The most likely way for such an allele to become prevalent would be for it to drift to a high frequency in a local population. Then the gene could spread by this local population's growing faster than others and ultimately replacing them, as in Wright's shifting-balance model.

Selection for an Altruistic Trait. How natural selection develops a trait that is harmful to the individual but helpful to others is not immediately obvious. How can such individuals increase? Yet such traits are known; for example, self-sacrificing behavior is frequently found.

The most striking example of self-sacrifice is the worker bee, who often dies in the process of stinging an animal that threatens the hive. This behavior is understandable because the worker bee, being sterile, can perpetuate her genes only by protecting the queen, with whom she shares half her genes. That is, those alleles that cause a worker to protect the queen are more likely to be perpetuated because the queen, being protected, will produce more progeny.

This viewpoint explains the evolution of complex cooperative behavior patterns in organisms with sterile classes, such as social insects. Similar theory has been used very effectively to explain several behavior patterns in bees; for example, why fertile workers (which arise occasionally) do not replace other workers. However, the bee example does not help to explain altruism in organisms without sterile castes.

We can look at this situation with another example. Genes that cause a mother to protect her progeny, even at some risk to herself, can increase in the population, since keeping the progeny alive perpetuates some of these same genes. Since an offspring shares 50 percent of her genes, it is worth a considerable risk (up to 50 percent, ideally) to keep the progeny alive. The risk taken might be greater if the mother were near the end of her reproductive period and the progeny were just reaching sexual maturity. We would expect natural selection to favor such traits, whether behavioral or physiological (such as one organism producing a chemical by-product useful to others).

We can make this verbal argument quantitative as follows. Suppose that an altruistic act entails a cost, c, measured as reduced fitness; the individual that helps another does so at some risk to itself. The benefit to the recipient, measured as increased fitness, is b. The numbers, c and b, are both positive and small relative to 1. Now suppose that the individual dispenses help to a relative that shares a fraction r of its genes. An allele or group of alleles that causes this pattern of behavior will tend to increase if $c < br$ ($r_{IJ} = 2F_{IJ}$ as was discussed in Chapter 5). We can write the general condition for a heritable altruistic trait to increase as

$$\frac{c}{b} < r, \qquad \text{7-16}$$

where r is the coefficient of relationship.

I know you realize that I do not intend to imply that animals compute kinship coefficients, only that natural selection would tend to increase any genes that caused a behavior pattern that approximates what an animal would do if it did make such a calculation. Protective behavior toward sibs, and especially toward progeny, is often seen in nature, and the **kin selection** just described is a sufficient explanation, although not a necessary one.

As an example of kin selection for a nonbehavioral trait, Fisher (1930; 1958, pages 177 ff) long ago suggested that this form of selection might explain the evolution of distastefulness of some butterfly larvae. A bird will learn which larvae taste bad by experimentation, and such an experiment is likely to be fatal to the larva involved. However, if the eggs have been laid in a cluster, a group of sibs will be close together.

The first larva is sacrificed while the bird learns, but others are saved. If more than two sibs are saved by such a process, the alleles causing the distastefulness will tend to increase. Fisher went further to suggest that a similar principle might explain the origin of self-sacrificing heroism in humans.

The intuitive derivation that I have given here can be replaced by rigorous mathematical derivations. An example is given by Aoki (1981). The principle was discovered by Haldane and Fisher, as usual, but formally developed by Hamilton (1964).

We are also interested in models that are somewhat similar but do not depend on kin recognition. In a small population all the individuals are eventually related to one other. Therefore alleles that cause altruistic behavior toward members of the group would tend to increase by the principles of kin selection.

The condition for such traits to increase in a subdivided population, assuming that they are heritable and that the requisite genetic variability exists, can be given in terms of G_{ST} (recall Chapter 3 and Table 3-2). The condition is

$$\frac{c}{b} < \frac{2G_{ST}}{(1 + G_{ST})} = r. \qquad \text{7-17}$$

By using isozyme markers, we can measure G_{ST} in natural populations and in this way determine what ratio of cost to benefit would tend to be selected. At last we have a justification for some of the calculations in Chapter 3. From Table 3-2 the average value of r is about 0.13. This value suggests that, provided the requisite genes exist and the population structure has been maintained for many generations, these primates would tend to evolve such that an individual would run a risk of up to 13 percent of the expected benefit to the recipient if the recipient were a member of the same troop. For a derivation and further discussion of this principle, see Crow and Aoki (1982, 1984).

Since until recently our ancestors had a population comprising many partially isolated tribes, such selection as I have been describing is a real possibility. Perhaps much of human social behavior — friendship, cooperation, and even some heroism and self-sacrifice — owes its existence to group population structure in our evolutionary past. How important this selection is, relative to all the other biological, social, and psychological factors that have influenced the evolution of human behavior, is still unknown.

Evolutionary Adjustment of Mutation Rates. We can reasonably expect that there is an optimum mutation rate. If the rate is too high, too many harmful mutations occur and the species or strain loses fitness and may become extinct in competition with others. On the other hand, if the rate is too low, the genetic variability may not be sufficient to keep up with a changing environment. I have repeatedly emphasized in this book the variance-conserving capacity of Mendelian inheritance and that a low mutational input is sufficient to maintain substantial genetic variance. Actually most species appear to have, if anything, more genetic variance than they seem to need, but we have no way of confirming this idea.

How can mutation rates be adjusted? Many physiological mechanisms will alter mutation rates. Numerous enzymes can repair different kinds of mutational damage, and various polymerases have different fidelities. There are several proof-reading mechanisms. In fact, by altering the ratio of exonuclease to polymerase activity of a DNA-replicating enzyme, the process can be changed from fast and error-prone to slow and accurate, or vice versa. These mechanisms work on the entire genome. Also, the mutation rates of different normal alleles vary at the same locus. Thus there are mechanisms for adjusting the mutation rate, both locally and globally.

The problem is how natural selection acts on mutation rates. In an asexual species, the method is relatively clear. A genotype with a lower mutation rate would have a selective advantage by producing a larger fraction of normal descendants. Presumably the rate would continue to be lowered by selection until it reached some sort of physical limit or until the mutation rate became too low to permit keeping up with a changing environment.

In a sexual species the problem is different. Genotypes do not last, so that the genes that alter the mutation rates soon become separated from the genes whose rates they adjust. Thus selection will be much less efficient. Individual alleles with lower mutation rates might be selected for, but here the selection is very weak—the difference between the two mutation rates. The alleles are likely to be selected for other differences more important than their mutation rates. For all these reasons mutational adjustment in a sexual species appears difficult. The mutation rate is quite possibly higher than optimum, since error-reducing mechanisms must be selected for.

One possibility is that the germinal mutation rate is actually a by-product of adjustment of somatic mutation rates. An organism with too high a somatic rate would clearly be at a disadvantage, with too many dead cells or malignancies. Selection for reduction of somatic mutation rate would be direct. Therefore one possibility is that, since presumably the same mechanisms affect both somatic and germinal rates, selection for reduced somatic mutation keeps the germinal rate down.

As is clear from this discussion, we know very little about the evolution of mutation rates. We shall have to await future analyses. However, one species whose mutation rate is clearly too high (at least in my opinion) is *Homo sapiens*. It is possible that as our life cycle became longer in the recent evolutionary past, the mutation rate adjustment, which is very slow at best, has not kept pace.

In the short-term future, say a dozen generations, we would clearly be better off if the mutation rate were 0. If someone discovers a safe way to reduce the human mutation rate to 0, or by any amount, I am for it. We have such a large store of existing genetic variability that mutation could stop entirely without our knowing it, except for the absence of some of our most loathsome dominant diseases. There is no danger of our running out of variability in the foreseeable future; there is enough variation for even the most wild-eyed eugenicist! And, should we ever at some time in the distant future need mutations, we surely know how to produce them.

7-5 EVOLUTIONARY ADVANTAGES OF SEXUAL REPRODUCTION

Sexual reproduction has many disadvantages. First, as a purely multiplying device, meiosis and fertilization are not especially efficient and many asexual systems are more effective. If there are separate sexes, two are needed to do the work of one; if females could reproduce parthenogenetically, the system would appear to be better. Furthermore, as we saw in Chapters 4 and 5, asexual selection acts on the total genetic variance rather than on just the genic or additive component, and therefore it should be correspondingly faster. Finally, the multiple-peak problem, discussed in connection with Sewall Wright's shifting-balance theory, does not exist in an asexual population. Whichever genotype is best, once it occurs, is perpetuated and has no valleys to cross.

Yet despite these drawbacks, Mendelian reproduction is found throughout nature. What are its advantages? What do segregation and recombination provide that makes the system so advantageous that it has become widespread despite the obvious defects? This is one more of the basic evolutionary questions whose answer is not known, or at least not fully understood. Several ideas have been proposed, and I shall mention a few that seem to me to be the most likely to be important. They are in three categories.

1. Putting Together Rare Beneficial Mutations. The first idea was put forth mainly by H. J. Muller, although Fisher proposed something rather similar. The dilemma, in this view, is that in an asexual population beneficial mutations that occur in different lineages can never be incorporated into the same individual. If mating and recombination occur, however, the two favorable genes can be brought into the same individual. Muller illustrated this principle through a famous diagram, which has been reproduced in many books. It is shown here as Figure 7-5.

The diagram shows only favorable mutations; harmful ones are quickly eliminated by natural selection. As the diagram depicts, favorable mutations that arise in separate lineages cannot merge in an asexual system. The only way in which two favorable mutations can be incorporated into the same individual is if the second mutation occurs in a descendant of the first. This situation may require a long time period, for the descendants of the first mutation must grow to a large enough number to make the second mutation probable. As we saw in Chapter 4, even if the first mutation is quite favorable, this growth takes a long time.

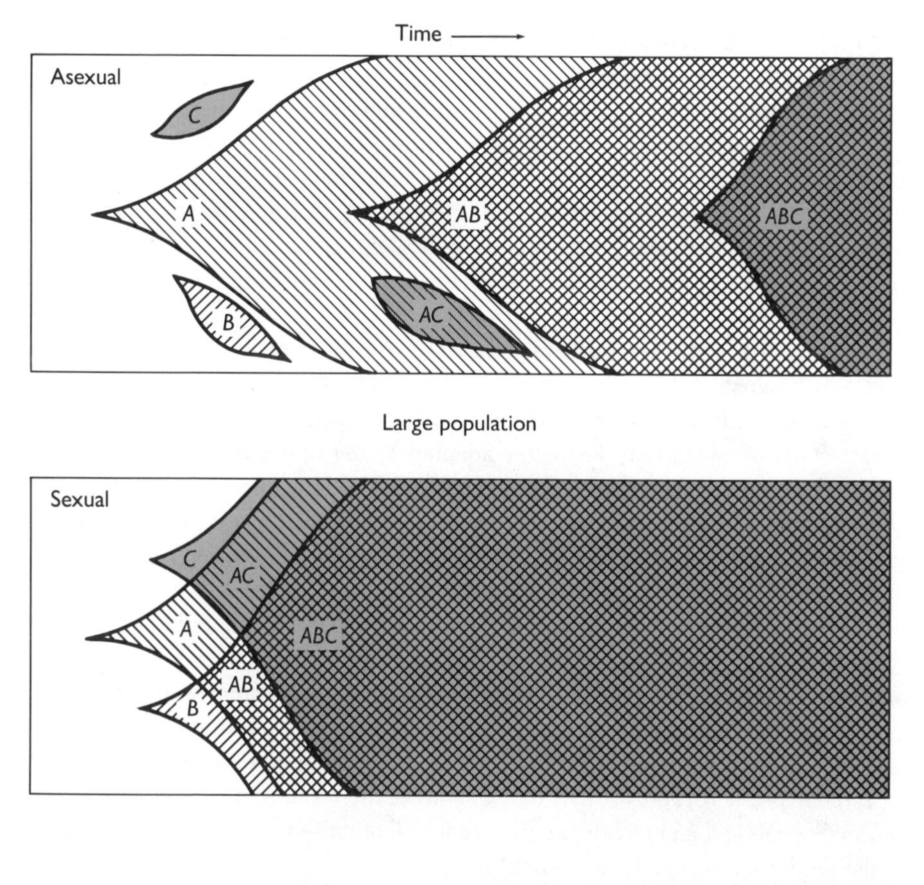

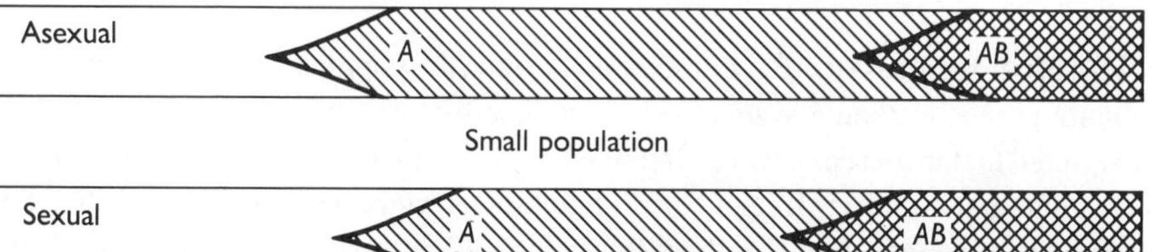

FIGURE 7-5. Incorporation of new, beneficial mutations in an asexual and in a sexual population. In an asexual population mutants that arose in separate individuals cannot get into the same individual. (Redrawn from J. Crow and M. Kimura. 1965. *Amer. Natur.* 99: 439; adapted from H. J. Muller. 1932. *Amer. Natur.* 68: 118.)

In contrast, in a sexual population the mutations that arise in separate individuals can recombine. The population does not have to wait for the descendants of the first mutant to become numerous. The larger the number of favorable mutations that occur at about the same time, the greater is the advantage of recombination. This relationship means that the advantage is greatest in a large population. The lower diagram shows that the advantage of recombination is less when the population is small, for the second mutation may be long delayed.

The principal criticism of the Muller theory is that many geneticists do not believe that putting together new mutations is the limiting factor in evolution. If, as has been suggested in earlier chapters, most responses to a changing environment involve recombining existing variants rather than incorporating rare, favorable mutations, then the Muller principle is not very important. It might, however, have been important in the early history of life, when recombination was first evolving.

2. Keeping Up with a Changing Environment. I mentioned earlier that a Mendelian population has a maximum of potential variance with a minimum of standing variance. It can respond to changes in the environment by producing recombinants, some of which may be better adapted to the new environment than any member of the current population. An asexual species, if it is to be prepared for large environmental changes, must be highly variable and hence has a large load of deleterious types.

We saw in the selection experiments in Chapter 4 that a few generations of selection can produce types well outside the range of the original population. It might be argued that changes in nature are not so drastic as to require such types; certainly evolutionary changes in size are not more than a tiny fraction as rapid as the changes in mouse size under human selection. Yet we must remember that fitness is multidimensional. An environmental change may change the optimum size, liver function, foot shape, blood pressure, and so on. Although none of these changes is outside the range of normal variability, an individual optimum for all of these traits may not exist in the population. Hence recombination could be advantageous.

3. The Mutation Load. Recombination permits the population to cope with recurrent mutation better than a system without it. The first way is by "Muller's ratchet." Muller noted that in an asexual population there is always the chance that no individual in the population is entirely free of harmful mutations. Once a harmful allele is fixed, it cannot be eliminated except by back-mutation. When one harmful allele is fixed, another may be at another locus, and the ratchet makes another turn. In a sexual population, in contrast, recombination can reconstitute the mutant-free type.

The second advantage of a sexual population is that it can remove harmful mutations in bunches and hence reduce the mutation load. Truncation selection is particularly effective in eliminating several harmful mutant alleles in one extinction, or "genetic death." In an asexual population the mutants stay in the lineage in which they occur; if they occur independently, as they surely do to a first approximation, then they are eliminated independently.

Which of these three advantages—(1) the ability to put together favorable mutations that occur in different individuals, (2) the ability to produce recombinant types that can keep up with a changing environment, or (3) the ability to get rid of or cope with harmful mutations—is most important is anybody's guess. Perhaps all three have played a significant role.

Many other ideas about the advantages of Mendelian reproduction have been suggested by evolutionists and ecologists. Most of these theories emphasize aspects other than recombination and therefore do not address the genetically important point.

I have discussed only what the advantages of recombination might be, not how it might be established and preserved by natural selection. Perhaps, as Fisher suggested, selection between groups is responsible. Those populations with recombination outcompete those without. Yet most theorists prefer individual selection to group selection. The whole subject is unresolved, and I am inclined to agree with J. Maynard Smith, who suggested that perhaps the crucial idea has not yet emerged.

The literature on this subject is extensive. If you would like to expand on this all too brief and perfunctory discussion, the books by Maynard Smith (1978) and Bell (1982) are good sources.

7-6 SPECIATION AND EVOLUTION OF HIGHER CATEGORIES

The neo-Darwinian paradigm is that the same processes that lead to minor changes in evolutionarily brief times lead to large changes in long times. Darwin certainly believed this. The idea has been mainly developed by students of evolution who approached evolution from the viewpoint of population genetics, which itself is primarily an extension of Mendelian inheritance. This view has become standard for those who study evolutionary mechanisms. Molecular studies have done little to support or to counter this view, much as they have enriched our understanding of basic mechanisms of inheritance and behavior of DNA. Paleontologists, on the contrary, have sometimes invoked other explanations.

I shall address two major questions. One is how species originate. The second is whether we need to postulate macroevolutionary phenomena, different from the microevolutionary factors discussed in this book, to explain evolution of higher categories, such as orders, classes, and phyla.

The origin of species involves not only the splitting of a single population into two, or possibly more, but preventing these two from hybridizing and becoming a single population again. Some sort of **isolating mechanism** is necessary. The existence of such mechanisms in closely related species is well documented. Matings between such species routinely produce offspring that are less viable and fertile than their parents. Often they are completely sterile; a familiar example is the mule. Sometimes only one sex is sterile or inviable, which is usually the male. The exception is in species in which the XY type is female; in these cases the females are more often the weaker or sterile hybrids. This tendency for hybrids of the heterogametic sex to be sterile is called

Haldane's rule, which is traditionally explained by the scarcity of genes on the Y chromosome. In female hybrids combinations of X-linked and autosomal genes of each species are preserved; in males they are not. This rule argues for the importance of coadaptation of the genes in each species.

In addition, there are usually strong mating preferences. Intraspecies matings are much more common than those between species when the individuals are given a choice. Natural selection would obviously favor such behavior in cases in which the hybrids are sterile or inviable. A major blunder, from the standpoint of natural selection, is to participate in a mating that would lead to inviable or sterile progeny.

The most widely accepted view is that speciation is **allopatric.** A population splits into two or more parts for some reason, often a geographical barrier, and the parts remain isolated from each other. Each population evolves independently of the other and gradually accumulates enough genetic differences to render the hybrids sterile or inviable. The process is accelerated if there are environmental differences in the various populations, but it will also eventually occur even if the isolated populations have similar environments. Perhaps it will happen almost as fast, given the innumerable genetically different ways of responding to the same environmental challenges. Speciation, that is, the divergence and development of mechanisms that prevent successful hybridization, is thus a by-product of long geographical isolation. If the geographical barrier later breaks down, the groups remain separate species because of the absence of hybridization and gene flow between them.

This mechanism of speciation is, I think, universally accepted. The point of controversy is whether or not **sympatric** speciation (speciation without geographical separation) *ever* occurs. At least one mechanism of sympatric speciation is well understood: polyploidy. A tetraploid that arises in a diploid population is immediately isolated, since hybrids with diploids would be sterile triploids. Do other mechanisms of sympatric speciation occur with any frequency? The issue is still unresolved. We know that allopatric speciation is common; whether or not sympatric speciation is of much evolutionary importance remains to be shown.

About 50 years ago Richard Goldschmidt argued that differences among species and among higher categories were too profound to have occurred by a succession of small steps, each of which was adaptive. How do you go from a two-winged to a four-winged insect by easy stages when an imperfect wing seems worse than none at all? Many people pondered this question, which had been raised ever since Darwin's book. But the overwhelming opinion was that mutants with such large effects—systemic mutations, or "hopeful monsters," as they were called—stretched credulity more than a succession of intermediate stages did. Monsters that deviate greatly from the normal strain certainly occur, but they are always weak or sterile and not promising candidates for future evolution. Drosophila geneticists have combined a small number of mutants to make a fly with four wings instead of two; but the innervation and appropriate musculature are not there, so that the insect cannot fly.

A Goldschmidtian view has recently been resurrected to explain the "punctuated equilibrium" mentioned earlier in connection with the seeming jerkiness implied by the fossil record. Most population geneticists doubt that any such assumption is required. What looks like an instant in paleontological time may be a large number of generations, and evolutionary gradualism would appear to suffice. But whether or not there are other, higher-level mechanisms and rules causally independent of those observed in microevolutionary studies remains a question for the future. Since the neo-Darwinian mechanism is sufficient, it seems, to me at least, best to wait for more convincing evidence before invoking macromutational hypotheses.

QUESTIONS AND PROBLEMS

1. In comparison of carp alpha and human beta hemoglobin, out of a total of 139 sites, 61 sites are the same, 49 differ by a minimum of a single step, and 29 require at least two mutations. Based on the 61 sites with no change, we estimated from Equation 7-1 that the mean number of replacements per site was 0.824. How does this value compare with the minimum number based on the additional data given here?

2. How might you expect the rate of evolution to compare for pseudogenes, noncoding DNA in general, and synonymous changes?

3. As stated in Chapter 2, the time required for a new mutant to go to fixation, assuming that it is successful, is $4N_e$. (a) Would you guess the time to be greater or less if the mutant is favored? (b) What would be your guess if the mutant is slightly unfavorable but drifts to fixation anyhow?

4. What would be the consequences of a segregation-distorting gene on the Y chromosome?

5. Suppose there are 10 loci, each with two semidominant alleles, and each allele contributes 1 inch to the height of a corn plant. What proportion of plants would be 20 inches above the average if each allele has a frequency 1/2?

6. On the hypothesis presented in this chapter, what would be the effect of female infanticide on the sex ratio at birth?

7. It is said that Haldane was once asked whether or not he would follow the biblical injunction and lay down his life for his brother. He is reported to have said, "No, but I would consider it for two brothers." How would you explain his answer?

8. Equation 2-13 predicts virtually complete heterozygosity for a neutral locus in a large population. How might you reconcile that average heterozygosity is only about 10 percent for isozymes with neutrality, even in large populations?

APPENDIX

PROBABILITY AND STATISTICS

I PROBABILITY

Definition of Probability. We start by letting E stand for some event, usually a future event. If n is the number of equally likely outcomes of which m are those in which E happens, then the probability of event E is

$$P(E) = \frac{m}{n}. \qquad \text{A-1}a$$

For example, let us designate event E as a 2 on a die. When the die is tossed, six equally likely outcomes are possible, so that the probability of a 2 is $1/6$. Or, the event E might be an odd number of spots, in which case the probability is $3/6$, since three odd outcomes are possible — 1, 3, and 5 spots. As another example event E might be the birth of an Aa child from two Aa parents. We know from the rules of Mendelian inheritance that there are four equally likely outcomes: AA, Aa, aA, and aa. The probability of a heterozygous child (we do not distinguish between Aa and aA) is 2/4, or 1/2. Likewise the probability of a homozygous aa child is 1/4. For still another example consider a deck of cards. The probability of drawing an ace from a deck (thoroughly shuffled) is 4/52, since there are 4 aces in the deck of 52 cards and each card is equally likely to be drawn.

Note that this definition implies that an impossible event (no outcome favorable, $m = 0$) has a probability of 0 and a certain event (all outcomes favorable, $m = n$) has a probability of 1.

This definition is applicable to situations in which we understand the mechanism well enough to know that the different outcomes are equally likely. Mendelian inheritance provides an example. In normal meiosis the two alleles from a heterozygote are equally likely to be included in a gamete. Likewise in the dice and card examples, we understand the mechanism well enough to adopt the definition represented by Equation A-1*a*.

We also frequently use empirical probabilities — probabilities based on experience. For example, about 130 babies per million births are albino. We can then say that the probability of an albino child is 13/100,000, or 1 in 7700. Our confidence in this definition increases as the amount of experience increases, which suggests a definition of empirical probability as the limit of m/n as n increases:

$$P(E) = \lim_{n \to \infty} \frac{m}{n}. \qquad \text{A-1}b$$

Alternatively we can describe probability subjectively. We can say that it represents our "degree of belief" in the truth of some statement. We surely believe that the probability of the sun rising tomorrow is high, considerably higher than the probability that the next president of the United States will be a Republican. But to quantify such a statement, we have to think of some sort of ratio of frequencies that specifies the betting odds by which to measure or describe our degree of belief.

You may find this definition unsatisfactory because of the circularity of defining probability by an expression that includes the term *likely*, which has essentially the same meaning as the word that we are trying to define. Mathematicians define probability axiomatically in terms of sets of points in a defined "sample space." This definition leads to easily derived sets of rules for manipulating probabilities (see, for example, the book by Feller). Whether these rules apply to the real world is not a mathematical question. You may not care how probability is defined, but rather simply accept a set of rules worked out by mathematicians and assume that these rules work in the real world of biology. You will be in good company, for this is what most biologists do. For the purposes of this book, what definition you choose to accept makes little difference.

I shall make no attempt to be mathematically rigorous in what follows. Rather, I shall resort to intuitive, common-sense arguments and illustrate them with familiar, usually genetic, examples.

The Addition ("Either-Or") Rule. The probability that either event A or event B will occur, sometimes called the **union** of A and B, is

$$P(A \text{ or } B) = P(A) + P(B) - P(AB), \tag{A-2}$$

where $P(AB)$ is the probability of occurrence of both events.

For example, what is the probability that a card randomly drawn from a deck will be either an ace or a spade? By definition the probability of an ace is 4/52 and a spade 13/52. One card, the ace of spades, is both an ace and a spade. Altogether, from Equation A-2, the probability of either an ace or a spade is

$$P(\text{ace or spade}) = \frac{4}{52} + \frac{13}{52} - \frac{1}{52} = \frac{16}{52}.$$

The probability is readily verified by actually counting the cards in the deck: 16 are either an ace or a spade (or both). If we had simply added the probabilities of an ace and of a spade, we would have counted the ace of spades twice.

Much of the time in population genetics, we are studying events that are **mutually exclusive,** meaning that the occurrence of one event precludes the occurrence of the other. In this case $P(AB) = 0$. The extension to more than two events is straightforward, so that we can state the following general rule:

> **The probability of one or another of a set of mutually exclusive events is the sum of the probabilities of the single events.**

For example, the probability of getting an odd number of spots in a throw of a die is $1/6 + 1/6 + 1/6$, or $1/2$. (We obtained the same answer earlier, simply by noting that there are three such outcomes among the six possible.) The probability of a child of two Aa parents being of the dominant phenotype, that is, AA or Aa, is $1/4 + 2/4$, or $3/4$.

Here is a useful (and obvious) corollary: If P is the probability of occurrence of an event, the probability of its nonoccurrence is $1 - P$.

Conditional Probability. We use the symbol $P(B \mid A)$ to stand for the probability of event B given event A, or conditional on event A. We define conditional probability as

$$P(B \mid A) = \frac{P(AB)}{P(A)}, \qquad \text{A-3}$$

where $P(AB)$ is the probability of both A and B.

Suppose that a woman who is heterozygous for the X-linked recessive gene for hemophilia marries a normal man. Letting H and h stand for normal and hemophilic alleles on the X chromosome and letting Y stand for the Y chromosome, the mating is $Hh \times HY$. From the rules of X-linked inheritance, we know that the four equally likely outcomes (assuming a 1 : 1 sex ratio) are HH female, Hh female, HY male, and hY male.

The probability is 1/2 that the child is male and 1/4 that it is hemophilic. Let A stand for the event that the child is male and B for the event that it is hemophilic. Then the conditional probability that the child is hemophilic given that it is male is $P(B \mid A) = P(AB)/P(A) = (1/4)/(1/2) = 1/2$. Obviously the probability that the child is hemophilic given that it is female is 0.

Here is a second example. Suppose that two parents of genotype Aa, where a is an autosomal recessive mutant, have a child of normal phenotype. What is the probability that the child is heterozygous? The conditional probability of being heterozygous given that the child is dominant phenotype is $[(2/4)/(3/4)] = 2/3$.

One more example: the conditional probability of a 2 spot in a throw of a die given that the outcome was an even number of spots is $[(1/6)/(3/6)] = 1/3$.

The Multiplication ("Both-And") Rule. The probability that events A and B will both occur, sometimes called the **intersection** of A and B, is

$$P(A \text{ and } B) = P(AB) = P(A)P(B \mid A). \qquad \text{A-4}$$

We can derive this equation simply by rearranging Equation A-3.

For example, the probability of a hemophilic son in the preceding example is $P(A)P(B \mid A) = (1/2)(1/2) = 1/4$. The probability of a heterozygous daughter is also $(1/2)(1/2)$.

As another kind of example, we draw a card from a deck and, without replacing it, draw another card. What is the probability that both cards are aces? In the form of Equation A-4,

$$P(\text{both aces}) = P(\text{ace on first draw})P(\text{ace on second} \mid \text{ace on first}) = \left(\frac{4}{52}\right)\left(\frac{3}{51}\right).$$

The 3/51 in the second term comes from the fact that after removal of the first ace, 51 cards are left, of which 3 are aces.

If the two events are **independent,** then the occurrence of the first event has no influence on the probability of the second. In this case $P(B \mid A) = P(B)$, which leads to the multiplication rule for independent events: $P(AB) = P(A)P(B)$. The extension to more than two events is straightforward:

The probability that all of several independent events will occur is the product of the individual probabilities.

If a is the autosomal recessive allele for albinism and both parents are Aa, what is the probability that the first child is a normal son? Letting $P(A)$ stand for the probability that the child is male and $P(B)$ for the probability that the child is normal, we have a probability of (1/2)(3/4), since the two events are independent. (The X-Y pair is independent of the autosomal pair of chromosomes carrying the A and a alleles.)

As another example, with a 1 : 1 sex ratio the probability that a sibship of five children will all be boys is $(1/2)(1/2)(1/2)(1/2)(1/2) = (1/2)^5$.

Here is an example that uses both the multiplication and addition rules. What is the probability that in a sibship of five all will be of the same sex? The probability that all will be boys is 1/32, as in the preceding example. Likewise the probability that all will be girls is 1/32, again by the product rule. All the same sex means that the siblings are either all boys or all girls, and since all boys and all girls are mutually exclusive, we can add their probabilities. Hence the desired probability is 1/32 + 1/32, or 1/16.

Bayes' Theorem. Bayes' theorem is an application of the rules of conditional probability and may be stated formally as

$$P(A \mid B) = \frac{P(A)P(B \mid A)}{P(A)P(B \mid A) + P(\bar{A})P(B \mid \bar{A})}, \qquad \text{A-5}$$

where $\bar{A}$ means the nonoccurrence of event A and the other terms have the same meanings as before. We can derive the formula as follows:

(a) $P(AB) = P(A)P(B \mid A)$ — by the multiplication rule

(b) $P(AB) = P(B)P(A \mid B)$ — by the multiplication rule

(c) $P(A \mid B) = \dfrac{P(AB)}{P(B)}$ — rearranging (b)

(d) $P(A \mid B) = \dfrac{P(A)P(B \mid A)}{P(B)}$ — substituting (a) into (c)

(e)
$$\begin{aligned} P(B) &= P(AB) + P(\bar{A}B) \\ &= P(A)P(B \mid A) + P(\bar{A})P(B \mid \bar{A}). \end{aligned}$$

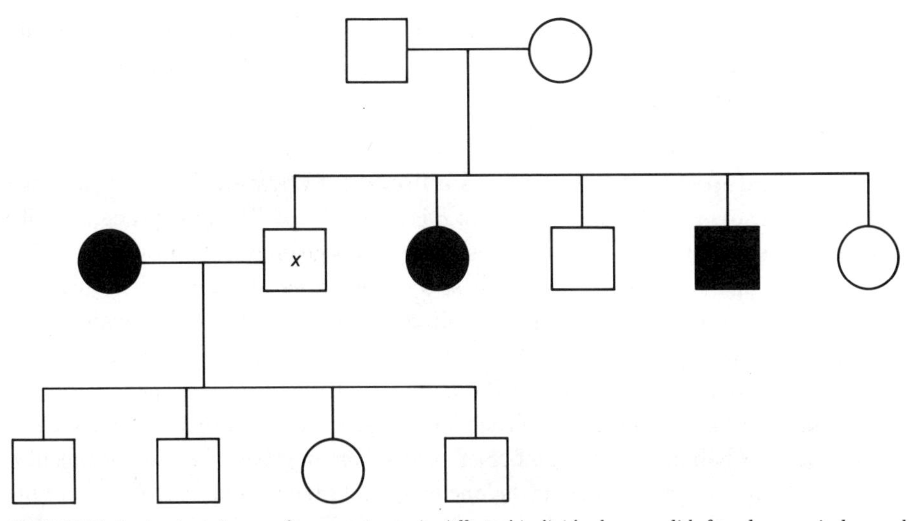

FIGURE A-1. A pedigree of a recessive trait. Affected individuals are solid; females are circles, and males are squares.

Statement (e) is true because B can occur in only two ways, with and without A. Substituting the right side of (e) for the denominator of (d) gives the expression in Equation A-5 that we are after.

Here is an example that illustrates the various complexities of conditional probabilities and Bayes' theorem. Consider the pedigree in Figure A-1. A normal man, designated by x, with two albino sibs (which tells us that his normal parents were both heterozygous for the recessive albino gene, a) marries an albino woman. What is the probability that the first child will be albino? The conditional probability that the man is heterozygous given that he has a normal phenotype is $2/3$, as computed in an earlier example. If he is heterozygous, the probability of an albino child is $1/2$. If we put these probabilities together, the probability that the first child will be albino is $(2/3)(1/2)$, or $1/3$. The probability that the first child will be normal is $(2/3)(1/2) + (1/3)(1) = 2/3$, as expected.

Now suppose this couple had four children and all were normal, as indicated in the figure. Each nonalbino child shifts the odds in favor of the father being of genotype AA. We can use Bayes' theorem to find out how much the odds have changed. Let A stand for the event that the father is Aa, $\bar{A}$ the event that he is homozygous normal AA, and B the event that all four children are normal. Then $P(A) = 2/3$, $P(\bar{A}) = 1/3$, $P(B \mid A) = (1/2)^4 = 1/16$, and $P(B \mid \bar{A}) = 1$, since if the father is homozygous, all the children will be normal. Substituting these numbers into Equation A-5 gives $P(A \mid B) = 1/9$. Having produced four normal children has shifted the probability of the father being heterozygous from $2/3$ to $1/9$. In probability jargon his prior probability of being heterozygous is $2/3$ and his posterior probability of being heterozygous is 1/9. If the parents have a fifth child, its chances of being albino are (1/9)(1/2), or 1/18, which is much less than the 1/3 it would have been if there had not been the four nonalbino children.

Repeated Independent Events — The Binomial Distribution. Assume that the probability of event A is p and of event B is q ($= 1 - p$). Then the probability of s occurrences of event A and t occurrences of event B is given by $p^s q^t$, provided that the events occur in a specified order, such as s occurrences of event A followed by t occurrences of event B. This example is a direct application of the multiplication rule.

However, we are usually more interested in asking the probability of a total of s occurrences of A and t occurrences of B without specifying the order. The way to solve this problem is direct. Since each order has the same probability, we have only to multiply the probability by the number of orders. The number of arrangements of n ($= s + t$) events, of which s belongs to one class and t belongs to another (where the events within a class are not distinguished), is given by $n!/s!\,t!$.

We can state the rule this way: If the probability of occurrence of event A is p and the probability of event B (or the nonoccurrence of A) is q, where $q = 1 - p$, then the probability that in n trials event A will occur a total of s times and event B will occur t times is

$$P = \frac{n!}{s!\,t!} p^s q^t, \qquad q = 1 - p. \qquad \text{A-6}$$

Notice that $s + t = n$ and $p + q = 1$. We also define $0! = 1$ and $X^0 = 1$; that is, any number raised to the 0 power is 1.

If the binomial expression $(p + q)^n$ is expanded, the individual terms in the expansion are those given by Equation A-6. Hence this distribution is often called the binomial distribution.

More Than Two Kinds of Events — The Multinomial Distribution. The extension of the binomial distribution to more than two kinds of events is straightforward. If the probabilities of events $A_1, A_2, A_3, \ldots$ are $p_1, p_2, p_3, \ldots$, the probability of n_1 occurrences of event A_1, n_2 occurrences of event A_2, and so on is

$$P = \frac{n!}{n_1!\, n_2\, n_3 \ldots} p_1^{n_1}\, p_2^{n_2}\, p_3^{n_3} \ldots, \qquad \text{A-7}$$

where $n = n_1 + n_2 + n_3 + \ldots$ and the p's add up to 1.

As an example assume again that two parents are each heterozygous for the recessive gene for albinism. The probability that a sibship of eight children will consist of four normal sons, three normal daughters, one albino daughter, and zero albino sons is

$$P = \frac{8!}{4!\,3!\,1!\,0!}\left(\frac{3}{8}\right)^4\left(\frac{3}{8}\right)^3\left(\frac{1}{8}\right)^1\left(\frac{1}{8}\right)^0 = 0.0365.$$

QUESTIONS AND PROBLEMS

1. Of the 148,462 deaths in New York in 1935, 4196 were from diabetes and 7436 were from tuberculosis. Based on these data, what is the probability that the first death in 1936 was from tuberculosis?
2. What is the probability that the first death was due to either diabetes or tuberculosis (assuming that a person cannot die of both)?
3. Assuming that a proportion p of newborns are boys and q are girls, what is the probability that a family of five children will consist of (a) all girls? (b) all the same sex? (c) three boys followed by two girls? (d) three boys and two girls in any order? (e) at least four girls? (If you would like a numerical answer, there are about 105.3 males born per 100 females; so $p = 0.513$.)
4. In one of Mendel's matings he obtained 315 round yellow seeds, 101 wrinkled yellow, 108 round green, and 32 wrinkled green out of a total of 556 seeds. The expectations are 9/16, 3/16, 3/16, and 1/16, respectively. What is the probability of obtaining exactly the same results as Mendel? (Do not bother to compute a numerical answer.)
5. A normal woman who had two sibs with Tay-Sachs disease (an autosomal recessive that is lethal in early childhood) marries a normal man who had a child with the disease in a previous marriage. (a) What is the probability that their first child will be affected? (b) If they have three normal children, what is the probability that the fourth will be affected?
6. An experimenter expects a 3 : 1 ratio of green to albino seedlings. The actual results are 80 green and 20 white among 100 plants. What is the probability of coming this close or closer to the expected number if the experiment is repeated? (Do not bother to compute the numerical answer. The next section will show how to get an approximate answer to problems of this type. You might give some thought as to how to answer the same kind of question for the data in Problem 4.)
7. Mendel studied seven traits. Also, the garden pea happens to have seven pairs of chromosomes. What is the probability that no two of the genes causing these traits are on the same chromosome?
8. There is a risk of hemolytic disease when the mother is Rh negative (*dd*) and the father is Rh positive (*DD* or *Dd*), since an Rh positive fetus has antigens to which

the mother may be capable of producing antibodies. Since usually two or more Rh positive fetuses are required before the mother produces sufficient antibodies, we want to determine whether the father is homozygous or heterozygous. In a marriage between an Rh negative woman and an Rh positive man, the first three children have been Rh positive. Knowing that in the United States white population, the conditional probability of a person being heterozygous given that he or she is Rh positive is $4/7$, what is the Bayesian probability that the husband is heterozygous?

9. (a) Assuming a $1:1$ sex ratio, what is the probability that two sibs will be of the same sex? (b) How much does this amount change if we introduce the correct sex ratio at birth, that is, $1.05:1.00$ (or 0.513 males)?

10. In the United States about $2/3$ of twin births are of the same sex. One-egg twins are always the same sex, whereas two-egg twins have the same probability of being the same sex as pairs of sibs do. What fraction of twins in the United States are monozygotic (from one egg)?

11. A pair of twins are the same sex. What is the probability that they are monozygotic?

12. Suppose every set of parents continues to have children until they have a son, and then they stop. Will this behavior alter the population sex ratio?

2 STATISTICAL MEASURES

One purpose of statistics is to describe succinctly the properties of a number of measurements too large for the properties to be comprehended from the raw data. At the minimum we need some idea of where the center of the distribution lies and how much variability there is. Sometimes additional measures that describe the general shape of the distribution are helpful, such as whether or not it is symmetrical about the central value.

Measures of Central Value. The most commonly used measure for the location of the center of the distribution is the familiar arithmetic average, or **mean.** The mean is the sum of the measurements divided by the number of measurements. Another sometimes useful measure is the **median,** the center value when the numbers are arranged in order from smallest to largest (or the average of the two central values if there is an even number of measurements). The median is often more appropriate than the mean if the distribution is highly asymmetrical or has more than one peak. For example, the median life expectancy may be closer to what we want to know than the mean life expectancy, especially in countries where a high rate of infant mortality makes the mean length considerably less than the median. Another useful measure is the **mode,** which is the measurement that occurs most frequently in the data. We conventionally describe distributions with two and more peaks as bimodal and multimodal.

It is mnemonically helpful to note that in most monomodal distributions the median is intermediate between the mean and the mode, just as it is when the words are alphabetically arranged. Furthermore the distance between mean and median is typically about half as great as the distance between median and mode; the two that are closest together are also nearest alphabetically.

Two other means that are frequently used in population biology are the harmonic and geometric means. The **harmonic mean** is the reciprocal of the average of the reciprocals of the numbers. It is useful when the quantity of importance exerts its influence in the denominator rather than the numerator. For example, random gene frequency drift is proportional to the reciprocal of the population number (see Section 2-4). Therefore the relevant average of a series of population numbers is the harmonic mean.

The **geometric mean** is the Nth root of the product of the N numbers. One way to compute this value is to average the logarithms of the numbers and then take the antilog. Biological data are often quite asymmetrical; for example, the distribution of weights is usually skewed toward the high side (that is, the longer tail of the distribution lies on the heavy side). The distribution of log weights is often nearly symmetrical, so that the geometric mean is more appropriate (for many purposes) than the arithmetic mean.

As an exercise consider the following hypothetical measurements (conveniently arranged in order of size): 6, 7, 8, 9, 9, 10, 10, 10, 11, 11, 12, 13, 17, 19, 21, 25, 30. You may wish to verify that the mode is 10, the median is 11, the mean is 13.41, the geometric mean is 12.12, and the harmonic mean is 11.10. Note that the arithmetic mean > geometric mean > harmonic mean. The harmonic mean is dominated by the smaller numbers, whereas the arithmetic mean is dominated by the larger ones. If the population size fluctuates, its effective number is dominated by a few bottlenecks.

Measures of Variability. One of the most widely used measures of variability is the **variance.** The variance is the average of the squared deviations from the mean (the mean is subtracted from each quantity, the differences are squared, and these squares are averaged). One difficulty is that the variance is measured in squared units: if the height is measured in centimeters, the variance is in square centimeters; if yield is measured in bushels, the variance is measured in squared bushels. To get around this difficulty, we frequently take the square root of the variance, which is called the **standard deviation.** The formulae for the calculation of some statistical measures are given in Table A-1.

TABLE A-I

Some standard statistical formulae. An individual measurement is designated by x or y. N_p is the population number, and N is the sample number. For the covariance, N or N_p is the number of pairs of measurements.

	POPULATION VALUE	ESTIMATE FROM A SAMPLE
Mean	$\mu = \frac{\Sigma x}{N_p}$	$m = \bar{x} = \frac{\Sigma x}{N}$
Variance	$\sigma_x^2 = \frac{\Sigma(x-\mu)^2}{N_p}$	$V_x = s_x^2 = \frac{\Sigma(x-\bar{x})^2}{N-1}$
Covariance	$\sigma_{xy}^2 = \frac{\Sigma(x-\mu_x)(y-\mu_y)}{N_p}$	$C_{xy} = \frac{\Sigma(x-\bar{x})(y-\bar{y})}{N-1}$
Standard deviation	σ_x	$s_x = \sqrt{V_x}$
Correlation	$\rho_{xy} = \frac{\sigma_{xy}^2}{\sigma_x\sigma_y}$	$r_{xy} = \frac{C_{xy}}{s_x s_y}$
Regression	$\beta_{yx} = \frac{\sigma_{xy}^2}{\sigma_x^2}$	$b_{yx} = \frac{C_{xy}}{V_x}$
Standard deviation of the mean		$s_{\bar{x}} = \frac{s_x}{\sqrt{N}}$

The difficulty with the standard deviation for many genetical and ecological problems is that standard deviations are not additive, whereas variances are. If x and y are independent measurements, then the variance of their sum is the sum of their variances. For example, if the height of a plant is the sum of a genetic component and an independent environmental component, the variance of height is the sum of the variances of the genetic and environmental components. For many situations additivity is more important than the appropriate dimensionality of the units. Furthermore selection changes a population in proportion to the genetic variance, not the genetic standard deviation. Thus the variance is often used instead of the standard deviation for the analysis of causes of variability.

We can use other measures of variability. One is the **average deviation,** the average of the absolute amounts by which each measurement differs from the mean. This measure seems sensible and it is sometimes preferred; for example, it is less sensitive to possibly irrelevant outlying observations (perhaps gross errors of measurement or of data recording) than is the variance, in which such deviations are squared. The reason why it is not used more often is that it is not naturally related to the normal distribution (see Section 3) the way the variance and standard deviation are.

Another measure of variability is the **interquartile range,** which is the difference between the cutoff points for the upper and lower 25 percent of the distribution. The distance between some other percentile cutoff values may be used. Such a measure is frequently useful for data that are not smoothly and symmetrically distributed.

In analogy to the moment of inertia in mechanics, the variance is often called the second moment. Statisticians often use higher moments to describe the distribution more fully. For example, the average of the cubes of the deviations gives a measure of the asymmetry , or **skewness,** of the distribution. However, we shall not use such measures in this book and shall concentrate on the mean and variance (or standard deviation). In most biological analyses the data are either assumed to be symmetrically distributed or have been transformed so that they are (for example, by using the logs, square roots, or some other transformation).

Sample and Population. Usually we are interested in a set of measurements only because they are thought to be representative of a larger universe. We collect a sample of mice from a population and would like to use the sample to draw inferences about the population or about mice in general. A quantity derived from measurements in a sample is called a **statistic.** The corresponding quantity in the population from which the sample was taken is called a **parameter.** Parameters are unknown, and we use statistics to estimate them. It is traditional to designate statistics by Roman letters and parameters by the corresponding Greek letters. For example, we use sigma (σ) for the unknown population standard deviation and s for the sample measurement that estimates sigma. (I shall bow to popular usage, however, and regularly designate the sample mean by $\bar{x}$ rather than by m.) Of course, the reliability of statistics as estimates of the corresponding parameters depends on the representativeness of the sample. The precision of the estimate depends on, among other things, the size of the sample.

The best estimate of the population mean is the sample mean, as you might expect. However, the best estimate of the population variance (best in the sense of having the correct expected value) is obtained by dividing the sum of the squared deviations not by the sample number N but by $N - 1$.

The formulae for computing these quantities are given in Table A-1. In these formulae an individual measurement is designated by x and the number of measurements by N, or N_p if the entire population is measured. If it is possible to observe the entire population, that is fine, but usually we have to be content with a sample. As expected, the variance of a series of means is less than the variance of individual measurements—in fact, it is $1/N$th as large. The standard deviation of the mean is then the standard deviation of individual items divided by the square root of the number in the sample.

In theoretical statistics it is important to keep the symbols for sample and population straight. In most practical procedures, such as those we are using, the terminological distinction is not carefully preserved. I shall follow popular custom and use the notation loosely where the distinction is not important. However, when variances are computed from sample measurements, we shall assume that the division has been by $N - 1$ rather than N.

Table A-2 gives some numerical illustrations. The individual numbers in the table were drawn from a population of numbers with a mean of 50 and a standard deviation of 10. Four samples of 30 were drawn, and the results are given in the table. (I have doctored the numbers slightly.) Notice that the sample means of the first four columns cluster around the expected value of 50, and the sample standard deviations are reasonably close to 10.

The numbers also illustrate the additivity of the variances. Notice that the variance of column 5, 375.68, is not far from the sum of the variances of columns 1, 2, 3, and 4, which is 397.38. However, the sum of the standard deviations is 39.78, almost twice as large as the standard deviation of the sum, which is 19.38; as we know, standard deviations are not additive.

The last column gives the means of the first four columns. Their standard deviation should be smaller than that of the individual measurements, and it is. The expected standard deviation of individual measurements is 10; that of the mean of four measurements is 10 divided by the square root of 4, or 5. The value at the bottom of column 8, 4.85, is in good agreement.

Covariance, Correlation, and Regression. The **covariance** is analogous to the variance. It is a measure of the similarity of two measurements, x and y. For example, x and y could be the height of two brothers. The covariance is the average of the products of the deviations of x and y from their means. The formulae are given in Table A-1. If x and y are the same, the covariance is the same as the variance. If x and y are independent, the expected covariance is 0, because the deviations of the individual values from their means will as often be in the opposite direction as in the same direction. Covariances, like variances, are additive.

It is convenient to convert the covariance into a dimensionless number whose value ranges from -1 to $+1$. The **coefficient of correlation** has this property. As Table A-1 shows, we obtain it by dividing the covariance by the product of the two standard deviations.

The coefficient of correlation can be interpreted as follows: if a measure is the sum of a series of equal components and if x and y share a fraction r of these components, while the remaining $1 - r$ are independent, then r is the expected correlation coefficient between x and y. This property is especially useful in genetics of quantitative traits, for example, in interpreting the correlations between relatives who share a certain fraction of their genes.

The **regression** of y on x, b_{yx}, is the slope of the straight line that best fits a set of pairs of measurements. For example, x could be the height and y be the weight of the same individual. The line that fits best is taken to be the line that minimizes the sum of

TABLE A-2

Illustration of some principles of variance and covariance. Columns 1 through 4 are random numbers from a population with mean = 50 and variance = 100. Column 5 = 1 + 2 + 3 + 4; column 6 = (1 + 2 + 3 + 4)/4.

	1	2	3	4	5	6
1	51	61	40	72	224	56.00
2	29	36	48	47	160	40.00
3	49	53	48	47	197	49.25
4	47	56	39	61	203	50.75
5	48	52	59	46	205	51.25
6	61	46	46	49	202	50.50
7	61	50	48	36	195	48.75
8	48	49	37	70	204	51.00
9	53	41	31	47	172	43.00
10	45	42	54	33	174	43.50
11	41	39	62	46	188	47.00
12	58	41	54	64	217	54.25
13	52	42	53	45	192	48.00
14	63	59	46	65	233	58.25
15	42	42	54	56	194	48.50
16	50	50	36	54	190	47.50
17	60	45	50	61	216	54.00
18	56	28	68	60	212	53.00
19	63	64	44	58	229	57.25
20	50	41	50	64	205	51.25
21	27	41	76	57	201	50.25
22	60	57	60	55	232	58.00
23	56	40	52	43	191	47.75
24	50	42	42	47	181	45.25
25	44	43	56	52	195	48.75
26	45	38	49	51	183	45.75
27	66	69	48	36	219	54.75
28	53	54	51	41	199	49.75
29	51	55	20	34	160	40.00
30	59	59	52	51	221	55.25
Total	1538	1435	1473	1548	5994	1498.50
$\bar{x}$	51.27	47.83	49.10	51.60	199.80	49.95
V_x	84.41	88.07	117.20	107.70	375.68	23.48
s_x	9.19	9.38	10.83	10.38	19.38	4.85

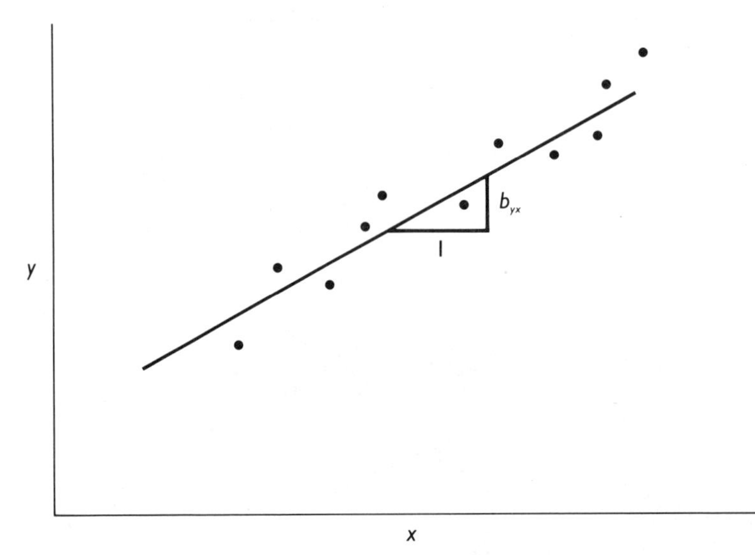

FIGURE A-2. The meaning of the regression coefficient. The line is drawn to minimize the sum of the squares of the vertical deviations from the data points. The regression coefficient, b_{yx}, is the slope of the line. This value is the amount that y changes when x changes by a unit amount.

the squares of the vertical deviations. The regression of y on x is the predicted amount by which y changes when x is changed by one unit. The regression coefficient is illustrated in Figure A-2. The regression of y on x is the covariance of x and y divided by the variance of x. The formula is given in Table A-1.

Variance of a Sum. Suppose a measurement, z, is the sum of two components, x and y. Thus we write $z = x + y$. Then the variance of z is given by

$$V_z = V_x + V_y + 2C_{xy}. \qquad \text{A-8}$$

If x and y are independent, then $C_{xy} = 0$, and the variance of a sum is the sum of the variances, as was illustrated in Table A-2.

We can easily derive Equation A-8 as follows:

$$\begin{aligned} V_{x+y} &= \frac{\Sigma[(x+y) - (\bar{x}+\bar{y})]^2}{N-1} \\ &= \frac{\Sigma[(x-\bar{x}) + (y-\bar{y})]^2}{N-1} \\ &= \frac{\Sigma(x-\bar{x})^2}{N-1} + \frac{2\Sigma(x-\bar{x})(y-\bar{y})}{N-1} + \frac{\Sigma(y-\bar{y})^2}{N-1} \\ &= V_x + 2C_{xy} + V_y. \end{aligned}$$

3
SOME USEFUL STATISTICAL DISTRIBUTIONS

The Binomial Distribution. We can think of the binomial distribution (see Equation A-6) as a series of outcomes in which each success is given a value 1 and each failure a value 0. If there are N trials and the probability of a success is p (more properly, we should use the Greek letter π if we are referring to the "true" value), the expected values of the mean and the variance are

$$\text{The mean number of successes} = Np$$
$$\text{The variance of the number of successes} = Np(1-p) \qquad \text{A-9}$$
$$\text{The variance of the proportion of successes} = \frac{p(1-p)}{N}.$$

The standard deviations are the corresponding square roots.

To derive these formulae, consider first a single trial and assign a value 1 to a success and a value 0 to a failure. The probability, or mean number of successes, is p. In N trials the mean is Np.

Now consider the variance of a single trial. From its definition the variance is $\Sigma[(x-\mu)^2 f(x)]/\Sigma f(x)$, where $f(x)$ is the expected frequency of x occurrences. If we apply this expression to a single trial, the variance is

$$(1-p)^2 p + (0-p)^2 q = pq(q+p) = pq,$$

where $q = 1 - p$. There is no denominator, since the two frequencies, p and q, add up to 1. The variance of a sum of a series of independent values is the sum of the variances; so the variance in the number of successes in N trials is $Np(1-p)$.

The proportion of successes is $1/N$ times the number of successes. Hence its variance is $(1/N)^2$ as great, or $p(1-p)/N$.

The Normal Distribution. If we let N get large, the binomial distribution approaches the *normal* (or Gaussian) distribution. Its equation is

$$y = P(x) = \frac{1}{\sigma\sqrt{2\pi}} e^{-(x-\mu)^2/2\sigma^2}, \qquad \text{A-10}$$

where μ $(=Np)$ is the mean and σ^2 $[=Np(1-p)]$ is the variance. It is not at all obvious why the limiting form of the binomial expression should contain π and e. The derivation, and hence the explanation, of this famous expression is beyond the scope of this appendix. However, the equation clearly indicates why statisticians measure variability by averaging the squares of the deviations rather than, for example, averaging the absolute values.

The curve has the familiar bell shape and it is completely determined by μ and σ. Because many sets of measurements are distributed approximately in this way (or can be transformed to be so), we can use the same equation either for sets of measurements or as a limit to the binomial distribution for enumeration data. It is convenient to scale the x values in units of the standard deviation, which has been done in Figure A-3.

The numbers along the x axis are scaled in units of the standard deviation, σ. The areas under the curve between two points on the abscissa are also given. For example, a fraction 0.341 of the observations is expected to lie between the mean and one standard deviation above the mean. We can also note that $2(0.341 + 0.136)$, or about 95 percent, of the population is within two standard deviations of the mean, or 99.8 percent (more accurately than the graph shows, 99.73) is within three standard deviations.

Almost all handbooks and statistics texts contain more extensive tables. You can also use the χ^2 and t chart on page 245. The quantity t is the ratio of the deviation from the mean or expected value to the standard deviation. Hold the chart so that it reads "Chart of t." Then look along the abscissa for the value of t, go up to the straight line diagonal corresponding to $N = \infty$, and then go to the left and read the probability. For example, if $t = 3$ (corresponding to a deviation three times the standard deviation), the probability is 0.003, the complement of the value 0.997 in the preceding paragraph. The t chart is arranged to give the total area in both tails of the distribution.

The curved lines on the chart are for the standard t test, which is used for measurement data when the sample sizes are small. A discussion of the t test is outside the scope of this book. You can find more information on this test in the statistics textbooks listed in the references. As for what the chart tells you when you hold it upside down, wait a few pages.

FIGURE A-3. The normal distribution curve. The mean and standard deviation are indicated by σ and μ. The numbers on the bottom indicate the fraction of the area under the curve within the brackets.

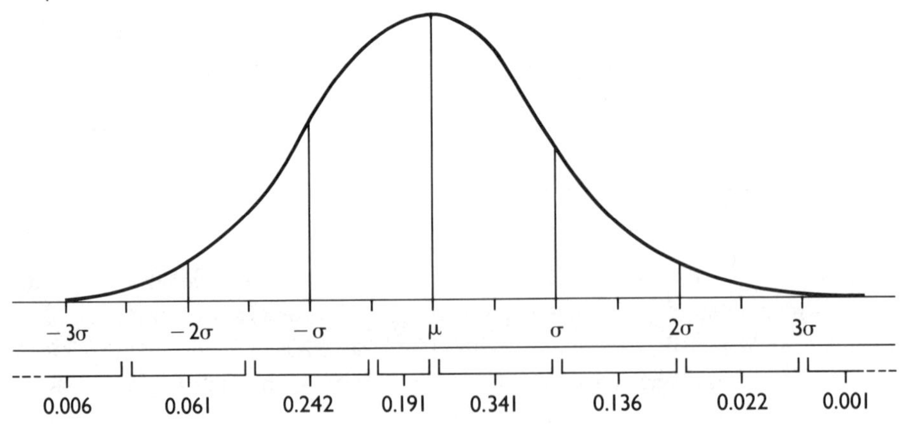

Example 1. Some traits that cannot be measured in any definable units, but which can be ranked, are often scaled to be normally distributed. For example, IQ scores are scaled to have a mean of 100 and a standard deviation of 15. The fraction of the population with IQs 130 or above is 0.023. Suppose a subset of the population has an average IQ of 107.5, or half a standard deviation higher than the population as a whole. What fraction of this group would have IQs 130 or above? In this case the deviation is 1.5 standard deviations, and looking at the graph in Figure A-3, we see that the proportion is 0.067.

Example 2. We can use the normal distribution to approximate the binomial distribution when the numbers are too large for easy calculation. In Problem 6 on page 221 an experimenter observed 80 green plants among a total of 100 when 75 were expected. What is the probability of obtaining, by chance, results that are this far or farther from the expected results? We note that when $N = 100$ and $p = 3/4$, the binomial variance (Npq) is 18.75 and the standard deviation is 4.33. The deviation from the expected result is $80 - 75$, or 5, and dividing this value by 4.33 gives 1.15. From the chart on page 245 a t value of 1.15 corresponds to a probability of 0.25. If this experiment were repeated 100 times, we would expect results this far or farther from the expected results 25 times. There is no reason to question the Mendelian assumption that led to these expectations.

Suppose the experimenter had observed 88 green plants instead of 80. Now the deviation from the expected result is $88 - 75$, or 13, and the t value is $13/4.33 = 3.0$. Looking again at the t chart, we see that the probability of deviating by this much in either direction is 0.003. Something seems to be wrong with the 3 : 1 expectation; for example, the albino plants may have been less viable. Either the expectation was wrong or an improbable event has happened.

Statistical Significance. By convention, if the probability is less than 0.05, the difference between observation and expectation is said to be **significant.** If the probability is less than 0.01, the difference is said to be **highly significant.** You should realize that these values are completely arbitrary, but they are useful for communication. When you read that a particular result is significant, you can interpret it as jargon for "the probability of getting a chance deviation from expectation this large or larger is less than 0.05." Notice that the t chart has the horizontal lines corresponding to probabilities of 0.05 and 0.01 thickened. You might want to write "highly significant" in the space above the 0.01 line (corresponding to probabilities less than 0.01), "significant" in the space between the two lines, and "not significant" below the 0.05 line.

The Chi-Square Distribution. This distribution is applicable to enumeration data when two or more categories are involved. It has the same relationship to the multinomial distribution as the normal distribution does to the binomial.

The value of chi-square is computed as

$$\chi^2 = \sum \left[\frac{(\text{observed number} - \text{expected number})^2}{\text{expected number}} \right], \qquad \text{A-11}$$

where the Greek capital sigma stands for summation. This value is then entered into a table or chart to give a probability. One other quantity is needed—the number of **degrees of freedom,** which is typically one less than the number of categories. We can best illustrate the use of this method with an example.

Table A-3 gives some data from a series of testcrosses in mink. The numbers expected from Mendelian theory are given in parentheses. The calculation of χ^2 is given below the numbers; because the number of categories is 4, the number of degrees of freedom is 3.

We obtain the probability by using the same chart as we used for the *t* test earlier, but turned upside down to give the chi-square test. Find the value 24.9 on the abscissa,

TABLE A-3

Data from a series of testcrosses in mink. The expected numbers, which are based on the assumption that the two loci are independent and that all classes survive equally well, are given in parentheses. The *E, e* and *B,b* loci determine coat color.

	Ee	*ee*	TOTAL
Bb	22 (20)	14 (20)	36 (40)
bb	7 (20)	37 (20)	44 (40)
Total	29 (40)	51 (40)	80 (80)

Testing *Bb* vs. *bb*: $\chi^2 = \frac{(36-40)^2}{40} + \frac{(44-40)^2}{40} = 0.8$

$df = 1;\ P = 0.37$

Testing *Ee* vs. *ee*: $\chi^2 = \frac{(29-40)^2}{40} + \frac{(51-50)^2}{40} = 6.1$

$df = 1;\ P = 0.015$

Testing both: $\chi^2 = \frac{(22-20)^2}{20} + \frac{(14-20)^2}{20} + \frac{(7-20)^2}{20} + \frac{(37-20)^2}{20} = 24.9$

$df = 3;\ P < 0.0001$

Data from R. Shackelford. 1949. *Amer. Natur.* 83:47.

go up to the line corresponding to $N = 3$, and then move to the left for the probability. In this case the value is off the chart, so the probability is less than 0.0001. Deviations of the observed numbers from their expectations this large or larger would be expected to occur by chance less than once in 10,000 similar experiments. Something is wrong with the hypothesis.

The χ^2 procedure gives only an approximation, but the approximation is to a quantity that would be very troublesome to compute directly. We would have to consider every possible outcome, compute their probabilities, and sum all those that give a probability less than that of the observed numbers.

To analyze the results further, we can examine the individual loci. Consider first the *B,b* locus. We observed 36 and 44 when 40 of each were expected, which leads to a $\chi^2 = 0.80$ with one degree of freedom. The probability from the chart is 0.37, not a significant deviation. Considering the *E,e* locus in the same way, we find a χ^2 value of 6.1, giving a probability of 0.015. This value is significant, probably indicating some inviability in the *Ee* mink. But the major discrepancy in the data is not in either of the two loci but in the relation between the two loci. We now consider this relation.

Test of Independence. To test independence in the 2×2 table, we treat the observed marginal (single locus) values as given and ask how much the internal values of the table deviate from their expectations if they were independent. The calculations are shown in Table A-4.

The expected numbers on the independence hypothesis are given in parentheses. If the two loci are independent, the internal numbers in the table will be proportional to the marginal values. The proportion in each category will be the product of the proportions in the two margins. For example, the expectation in the upper left corner is $80(36/80)(29/80) = 13.05$. We can compute the other three expected values in the same way or more easily by subtraction from the marginal totals.

TABLE A-4

The same data as were shown in Table A-3. The expected numbers, given in parentheses, are based on the single-locus totals given by the margins.

	Ee	ee	TOTAL
Bb	22(13.05)	14(22.95)	36
bb	7(15.95)	37(28.05)	44
Total	29	51	80

$\chi^2 = 17.5$
$df = 1$
$P < 0.0001$

Chi-square is computed in the same way as before. There are four terms that add up to 17.5. There is only one degree of freedom (not three as you might think) because as soon as we have assigned one expected value, the others are fixed by the necessity for the numbers in the table to add up to the marginal totals. A χ^2 of 17.5 with one degree of freedom corresponds to $P < 0.0001$. Clearly the two loci are not independent. (From your knowledge of genetics, you probably suspect that the reason for the nonindependence is that the two loci are located on the same chromosome pair.)

The general rule for degrees of freedom is as follows:

The number of degrees of freedom is the number of categories minus the number of *independent* ways in which the expected numbers are made to conform to the observed numbers.

In testing a 3 : 1, 1 : 1, or 1 : 1 : 1 : 1 ratio, we restricted the expected numbers in only one way—by making their total agree with the observed total. If n is the number of categories, there are $n - 1$ degrees of freedom. In a test of independence the expected numbers are made to agree with the observed numbers in two more ways, with the numbers in both margins. Two more degrees of freedom are lost, and the number is $4 - 3$, or 1. You can satisfy yourself that there is only one degree of freedom by inserting an arbitrary value in one of the squares. You will see that the other three values are then fixed by the marginal values; you had only one free choice.

Here are two helpful reminders about the χ^2 test. The first is to remember that you must use the observed and expected *numbers*, not their proportions. The second is that this procedure is only an approximation to the exact multinomial or other finite distribution. The approximation becomes poor when the numbers are too small. The usual rule of thumb is not to use the chi-square test when any expected number is less than 5.

The Poisson Distribution. One of the most commonly used distributions in genetics and many other sciences is the Poisson distribution. It is the limit of the binomial distribution as N gets large while, at the same time, p becomes small in such a way that the product Np remains finite. If we take this limit, the probability of any number x is

$$P(x) = \frac{e^{-\mu}\mu^x}{x!}, \qquad \mu = Np. \qquad \text{A-12}$$

Notice that since $Np = \mu$ remains finite but p approaches 0, $Np(1 - p)$ approaches Np, or μ. Thus the variance is equal to the mean, a very convenient property. Ecologists often use this principle to study the distribution of animals and plants over geographical areas. If the variance is less than the mean, the spacing is more uniform than random, as we would expect with territorial behavior. If the variance is greater than the mean, the animals or plants tend to cluster.

Example 1. Suppose a field of mice is randomly distributed with an average of 2 per quadrat. What proportion of quadrats would be expected to contain (a) 0, (b) 1, (c) 2, and (d) 3 or more mice? The answers are: (a) $e^{-2} = 0.1351$; (b) $e^{-2}(2) = 0.271$; (c) $e^{-2}(2^2)/2! = 0.271$; (d) $1 - 0.135 - 0.271 - 0.271 = 0.323$.

Example 2. In an experiment 1 ml of city water is put into each of 100 nutrient tubes. Forty of them showed bacterial growth, indicating that at least one bacterial cell was in the inoculum. What is the mean number of bacteria per milliliter? We answer the question by noting that $e^{-m} = 1 - 0.40 = 0.60$; $m = -\ln(0.60)$, or 0.51. We conclude that there is an average of about half a bacterium per milliliter.

QUESTIONS AND PROBLEMS

1. Humans and fish differ in 68 amino acids out of 141 in alpha hemoglobin. When they differ, there may have been undetected intermediate changes as they diverged during evolution. If changes occur at random, what is the mean number of changes per amino acid site? The ancestral lines of fish and humans diverged about 400 million years ago. Assuming that there have been 800 million years of independent evolution of these lines, what is the rate of amino acid substitution per year?
2. A series of infants who had congenital dislocation of the hip and who had a twin were selected. The results were as follows (concordant means that the other twin was also affected):

	CONCORDANT	DISCORDANT	TOTAL
One-egg twins	40	60	100
Two-egg twins (same sex)	20	80	100
Total	60	140	200

   Is the concordance independent of zygosity?
3. From an F_2 mating of sweet peas, $Pp\ Ll \times Pp\ Ll$ (where P stands for purple; p, red; L, long; and l, short, and the capital letters indicate dominant alleles), the progeny were 226 purple long, 95 purple short, 97 red long, and 1 red short. Test these values against a 9 : 3 : 3 : 1 expectation. Also test each of the single-locus ratios and the independence of the two loci.
4. Test by χ^2 the significance of the examples in Problems 4 and 6 on page 221.

4
STATISTICAL ESTIMATION BY MAXIMUM LIKELIHOOD

Sometimes the data gathered are only indirectly related to the population parameter being estimated. In such cases there are often several approaches to the problem, and it is not obvious which is best. One of the most widely used approaches is the method of maximum likelihood. R. A. Fisher showed that in large samples this method has a number of optimum properties, some of which I shall discuss later. It is perhaps best illustrated with an example.

Consider the F_2 linkage example in Problem 3 of the previous section. The mating is $P\ l/p\ L \times P\ l/p\ L$. If we let r stand for the proportion of recombinants between the two loci, we can calculate the expected frequency of the different F_2 progeny genotypes as shown in Table A-5. Collecting the expected proportions for each phenotypic class and simplifying the algebraic expressions leads to the following expected proportions:

		OBSERVED NUMBER	EXPECTED PROPORTION
Red, short	*pp ll*	$1 = D$	$r^2/4$
Red, long	*pp L–*	$97 = C$	$(1 - r^2)/4$
Purple, short	*P– ll*	$95 = B$	$(1 - r^2)/4$
Purple, long	*P– L–*	$226 = A$	$(2 + r^2)/4$
Total		$419 = N$	1

For convenience I shall let $x = r^2$. We now have four ways to estimate x (and therefore r) by taking the square root.

$$\frac{x}{4} = \frac{1}{419}, \quad \text{giving } x = 0.00955, \quad r = 0.098$$

$$\frac{1-x}{4} = \frac{97}{419}, \quad \text{giving } x = 0.07399, \quad r = 0.272$$

$$\frac{1-x}{4} = \frac{95}{419}, \quad \text{giving } x = 0.09308, \quad r = 0.305$$

$$\frac{2+x}{4} = \frac{226}{419}, \quad \text{giving } x = 0.15752, \quad r = 0.397.$$

What should we do now? We have four estimates, but it is far from obvious how to average them. The simple arithmetic mean is 0.268, but the numbers surely should not be weighted equally. The purple, long class has the largest numbers, but the

TABLE A-5

Expected proportions of the various genotyes in the mating *P l/p L* × *P l/p L*. The proportion of recombination between the two loci is *r*.

		SPERM			
		$P\ L$ $r/2$	$P\ l$ $(1-r)/2$	$p\ L$ $(1-r)/2$	$p\ l$ $r/2$
EGGS	$P\ L$ $r/2$	$PP\ LL$ $r^2/4$	$PP\ Ll$ $r(1-r)/4$	$Pp\ LL$ $r(1-r)/4$	$Pp\ Ll$ $r^2/4$
	$P\ l$ $(1-r)/2$	$PP\ Ll$ $r(1-r)/4$	$PP\ ll$ $(1-r)^2/4$	$Pp\ Ll$ $(1-r)^2/4$	$Pp\ ll$ $r(1-r)/4$
	$p\ L$ $(1-r)/2$	$Pp\ LL$ $r(1-r)/4$	$Pp\ Ll$ $(1-r)^2/4$	$pp\ LL$ $(1-r)^2/4$	$pp\ Ll$ $r(1-r)/4$
	$p\ l$ $r/2$	$Pp\ Ll$ $r^2/4$	$Pp\ ll$ $r(1-r)/4$	$pp\ Ll$ $r(1-r)/4$	$pp\ ll$ $r^2/4$

genotypes are not known. The red, short class has a known genotype, but there is only one plant and thus it has little statistical power.

The maximum likelihood method provides one widely used way out of the dilemma. In addition to its intuitive appeal, it has a number of optimum statistical properties, first demonstrated by R. A. Fisher.

The Maximum Likelihood Method. The logic of this method is simple: we ask for the value of r (or x) that maximizes the probability of the observed results. I think you will agree that this seems reasonable. This probability is

$$P = \frac{419!}{226!\ 95!\ 97!\ 1!}\left(\frac{1}{4}\right)^{419}(2+x)^{226}(1-x)^{95+97}x, \qquad \text{A-13}$$

and a plot of P is given in Figure A-4.

We can see from the graph that the value of r that maximizes the probability is about 0.12. But we want to be able to solve for the value algebraically. To do this, we follow the usual practice of equating the derivative of P with respect to x to 0. This leads to some troublesome algebra. But there is a simplification: we note that the value of x that maximizes the probability is the same value of x that maximizes the log of the probability. We can avoid a great deal of trouble by working with the log of the probability (that is, log to the base e). We also need not pay any attention to the various constant quantities, for their derivatives will be 0. We write

$$L = \ln P = A\ln(2+x) + (B+C)\ln(1-x) + D\ln x + \text{a constant.}$$

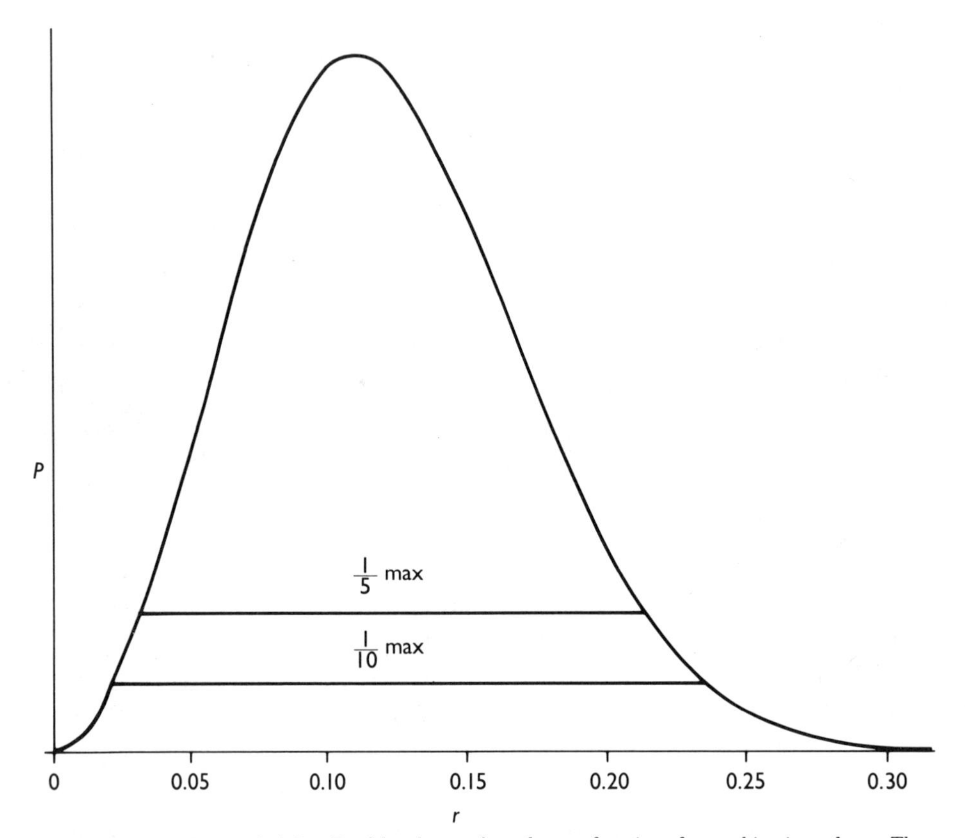

FIGURE A-4. The probability, P, of the observed results as a function of recombination value, r. The maximum of the curve is at about 0.12.

Equating the derivative to 0, we get

$$\frac{dL}{dx} = \frac{A}{2+x} - \frac{B+C}{1-x} + \frac{D}{x} = 0. \qquad \text{A-14}$$

Inserting the numerical values for A, B, C, and D and solving, we obtain

$$x = 0.01219 \qquad r = 0.1104.$$

We have used one property of maximum likelihood estimates, that of **functional invariance:** the maximum likelihood estimate of a function of a quantity is that same function of the maximum likelihood estimate of the quantity. In this case the maximum likelihood estimate of the square root of x is the square root of the maximum likelihood estimate of x. It is much easier to estimate x than r; so we estimate x and take the square root of the answer, thus using the functional invariance principle.

Testing by the Chi-Square Test. We can use the estimated value of r (or x) to compute the expected numbers for a test of significance. For example, the expected number of purple, long plants is $N(2+x)/4$, where $N = 419$; and similarly for the other classes.

	OBSERVED	EXPECTED	$\frac{\text{(OBSERVED-EXPECTED)}^2}{\text{EXPECTED}}$
Purple, long	226	210.8	1.10
Purple, short	95	103.5	0.70
Red, long	97	103.5	0.41
Red, short	1	1.3	0.07
Total	419	419.1	2.28

How many degrees of freedom are there? There are four categories, which suggests three degrees of freedom. But we have made the expected numbers conform to the data in another way, the value of r. So we lose another degree of freedom, and only two are left. A χ^2 of 2.28 with two degrees of freedom corresponds to a probability of 0.32; so there is good agreement with the expectations.

I have violated our rule of not using the χ^2 test when an expected number is less than 5. One way around this difficulty is to pool the two smallest classes. Thus suppose we add the red, short and purple, short classes, giving an observed number of 98 and an expected number of 104.8. We lose one degree of freedom for each such pooling. The contribution to chi-square from this pooled class is 0.44. Adding this quantity to 1.10 and 0.41 gives a χ^2 of 1.95 with one degree of freedom, or a probability of 0.16. The conclusion is unchanged.

The Variance of a Maximum Likelihood Estimate. Maximum likelihood theory provides a convenient way to compute the variance (and therefore the standard deviation) of an estimate. We calculate this value by differentiating Equation A-14 again, and then the negative reciprocal of the second derivative gives the variance

$$\frac{d^2L}{dx^2} = -\left[\frac{A}{(2+\hat{x})^2} + \frac{B+C}{(1-\hat{x})^2} + \frac{D}{\hat{x}^2}\right] = -\frac{1}{V_x}. \qquad \text{A-15}$$

The circumflex means that, after taking the derivative, we replace x by its maximum likelihood estimate, 0.01219. In this example $V_x = 0.0001129$.

We do not want the variance of x but rather the variance of its square root, r. Again, maximum likelihood theory comes to the rescue. There is a maximum likelihood theorem that says that for any x and y

$$V_y = V_x \left(\frac{dy}{dx}\right)^2. \qquad \text{A-16}$$

Since $x = r^2$, $(dr/dx)^2 = 1/4x$, and $V_r = V_x/4x = 0.002315$. The standard deviation of r is the square root of this value, or 0.048. We can say that the estimate of the recombination value is 0.110 with a standard deviation of 0.048. This error is rather large, considering that 419 progeny plants were counted. It shows that F_2 data are not very efficient for estimating recombination values.

Multiple Alleles. The maximum likelihood procedure can be extended to the simultaneous estimation of several parameters. A good example is the snail colors mentioned in Chapter 1.

The data are as follows:

COLOR	NUMBER	GENOTYPES	EXPECTED PROPORTION
Brown	$88 = A$	BB, Bb', Bb	$p^2 + 2pq + 2pr = 1 - x$
Pink	$83 = B$	$b'b', b'b$	$q^2 + 2qr = x - y$
Yellow	$42 = C$	bb	$r^2 = y$
Total	$213 = T$		1

I have let $y = r^2$ and $x = (q + r)^2$; $p + q + r = 1$. The three frequencies to estimate are p, q, and r, of which two are independent. It is more convenient to estimate x and y and use the principle of functional invariance.

The log of the probability of the observed results is

$$L = A \ln(1 - x) + B \ln(x - y) + C \ln y + \text{a constant}.$$

Now to find the values of x and y that maximize L, we equate the two partial derivatives to 0. Thus

$$\frac{\partial L}{\partial x} = -\frac{A}{1 - x} + \frac{B}{x - y} = 0$$

$$\frac{\partial L}{\partial y} = -\frac{B}{x - y} + \frac{C}{y} = 0.$$

These two equations are linear, and the solutions are

$$x = \frac{B + C}{T} = 0.587 \qquad y = \frac{C}{T} = 0.197,$$

from which $p = 0.234$, $q = 0.322$, and $r = 0.444$.

These values are the same as those we obtained by a more elementary procedure in Chapter 1; so the maximum likelihood method was not really needed. I used the example here as an illustration of the method. It also shows that the maximum likelihood method gives the common sense solution in this case, in which there *is* a common sense solution. To obtain standard errors for these estimates, see C&K, page 513.

An example without an obvious solution is provided by the ABO blood group data in Chapter 1. In this case the data are as follows:

GROUP	OBSERVED NUMBER	EXPECTED PROPORTION
A	$44 = A$	$p^2 + 2pr$
B	$27 = B$	$q^2 + 2qr$
AB	$4 = C$	$2pq$
O	$88 = D$	r^2

$$L = A \ln(p^2 + 2pr) + B \ln(q^2 + 2qr) + C \ln pq + D \ln r^2 + \text{a constant.}$$

We need to estimate three gene frequencies. Only two are independent, however; so we replace r by $1 - p - q$. Making this substitution and simplifying, we obtain

$$\begin{aligned} L = {} & (A + C)\ln p + (B + C)\ln q + A \ln(2 - p - 2q) \\ & + B \ln(2 - 2p - q) + 2D \ln(1 - p - q) + \text{a constant.} \end{aligned}$$

Taking the two partial derivatives, we have

$$\frac{\partial L}{\partial p} = \frac{A + C}{p} - \frac{A}{2 - p - 2q} - \frac{2B}{2 - 2p - q} - \frac{2D}{1 - p - q} = 0$$

$$\frac{\partial L}{\partial q} = \frac{B + C}{q} - \frac{2A}{2 - p - 2q} - \frac{B}{2 - 2p - q} - \frac{2D}{1 - p - q} = 0.$$

The only problem is that these equations are hard to solve. If you have several hours, you can do it by trial and error, perhaps in a shorter time if you make a good first guess. Every computing center has programs for solving maximum likelihood equations. There are systematic ways, developed by Fisher, for converging on the answer and getting the variance automatically (see Cavalli-Sforza and Bodmer, 1971, pages 885–888*). The estimates, with standard deviations in parentheses, are

$$p = 0.1605\ (0.0213)$$
$$q = 0.1004\ (0.0171)$$
$$r = 0.7392\ (0.0254).$$

The crude method in Chapter 1 gave almost identical answers; so in this case the more cumbersome maximum likelihood method was not worth the effort. We are not always so lucky, though, and the maximum likelihood is the better procedure.

A Final Word of Caution. The statistical significance tests and standard deviations given in this section are appropriate when the errors are due solely to sampling errors, caused by the smallness of the size of the sample. In the real world of biological research or the practical applications thereof, there are many other sources of error, such as errors of measurement, biological variability, and influences that are not taken into account in the experimental design. The best way to deal with these problems is to use an empirical measure of the error, based on repeated experiments. In good experimental design, replicates are built in to give an estimate of the error.

Many statistical tests depend on the assumption that the measurements are normally distributed. This assumption is often, but not always, true. Several tests do not demand this assumption; they are called nonparametric tests. You can find these methods in the texts listed in the references.

Be wary of conclusions based on theoretical variances or the assumption of normality. You may be well advised to examine the data more carefully.

QUESTIONS AND PROBLEMS

1. Show that for an ordinary genetic testcross (that is, $AB/ab \times ab/ab$), the maximum likelihood estimate of r is the same as the value you have always used.

2. In the F_2 linkage problem, if the amount of recombination is not the same in the two sexes, what does x estimate? Is it the mean of the recombination values in males and females?

3. In mice the genotype $B-\ C-$ is black, $bb\ C-$ is brown, $--\ cc$ is white (a dash

* My example came directly from this reference. In case you refer to it, let me correct a numerical error. The quantity 0.0000503, which appears twice in the covariance matrix on page 887, should have a negative sign. if you make this correction, you will discover that the values for p and q, correct to six decimal places, are achieved on the first iteration.

indicates either a capital or a lowercase letter). In a series of matings between $B\ C/b\ c$ and $b\ c/b\ c$, there were 81 black, 19 brown, and 90 white progeny. Estimate the amount of recombination between the B and C loci.

4. Use the maximum likelihood method to estimate the A and B allele frequencies under hypothesis 1 in the example on page 24 in Chapter 1. (Note that this method gives the same answer as the simple method used in the example.)
5. Use the maximum likelihood method to estimate the frequency of the color-blind gene in Table 1-10.

5
MATRIX TREATMENT OF AN AGE-STRUCTURED POPULATION

Equations 6-1 and 6-2 can be conveniently written in matrix form. The two equations, the second slightly rewritten, are

$$n_{xt} = n_{x-1,t-1} p_{x-1}$$
$$n_{0t} = n_{0,t-1} p_0 b_1 + n_{1,t-1} p_1 b_2 + \ldots + n_{k-1,t-1} p_{k-1} b_k.$$

We can write these equations in matrix form as

$$\begin{pmatrix} n_{0t} \\ n_{1t} \\ n_{2t} \\ \cdot \\ \cdot \\ \cdot \\ n_{kt} \end{pmatrix} = \begin{pmatrix} p_0 b_1 & p_1 b_2 & p_2 b_3 & \cdots & p_{k-1} b_1 & p_k 0 \\ p_0 & 0 & 0 & \cdots & 0 & 0 \\ 0 & p_1 & 0 & \cdots & 0 & 0 \\ \cdot & \cdot & \cdot & \cdots & \cdot & \cdot \\ \cdot & \cdot & \cdot & \cdots & \cdot & \cdot \\ \cdot & \cdot & \cdot & \cdots & \cdot & \cdot \\ 0 & 0 & 0 & \cdots & p_{k-1} & 0 \end{pmatrix} \begin{pmatrix} n_{0,t-1} \\ n_{1,t-1} \\ n_{2,t-1} \\ \cdot \\ \cdot \\ \cdot \\ n_{k,t-1} \end{pmatrix}$$

or, more concisely, as

$$\mathbf{n}_t = \mathbf{M}\ \mathbf{n}_{t-1}.$$

Likewise

$$\mathbf{n}_{t-1} = \mathbf{M}\ \mathbf{n}_{t-2}.$$

Continuing this process,

$$\mathbf{n}_t = \mathbf{M}\ \mathbf{n}_{t-1} = \mathbf{M}^2\ \mathbf{n}_{t-2} = \mathbf{M}^t\ \mathbf{n}_0.$$

The usual rules of matrix multiplication work, so that we can find the composition of the population at any time in the future by raising the matrix to the appropriate power. The column vectors are the age distributions. The matrix is usually called a **projection matrix.**

The asymptotic growth rate, λ, can also be found directly. We write the determinant as follows:

$$\begin{vmatrix} p_0b_1 - \lambda & p_1b_2 & p_2b_3 & \cdots & p_{k-1}b_k & p_k0 \\ p_0 & -\lambda & 0 & \cdots & 0 & 0 \\ 0 & p_1 & -\lambda & \cdots & 0 & 0 \\ \cdot & \cdot & \cdot & \cdots & \cdot & \cdot \\ \cdot & \cdot & \cdot & \cdots & \cdot & \cdot \\ \cdot & \cdot & \cdot & \cdots & \cdot & \cdot \\ 0 & 0 & 0 & \cdots & p_{k-1} & -\lambda \end{vmatrix} = 0.$$

This equation is solved for λ. It has only one positive real root, whose value is the rate of increase when age stability has been achieved.

The other roots and the eigenvectors also have uses and interpretations. For a discussion see Keyfitz (1968). For an introduction to matrix methods with applications to inbreeding, see C&K, pages 500–509.

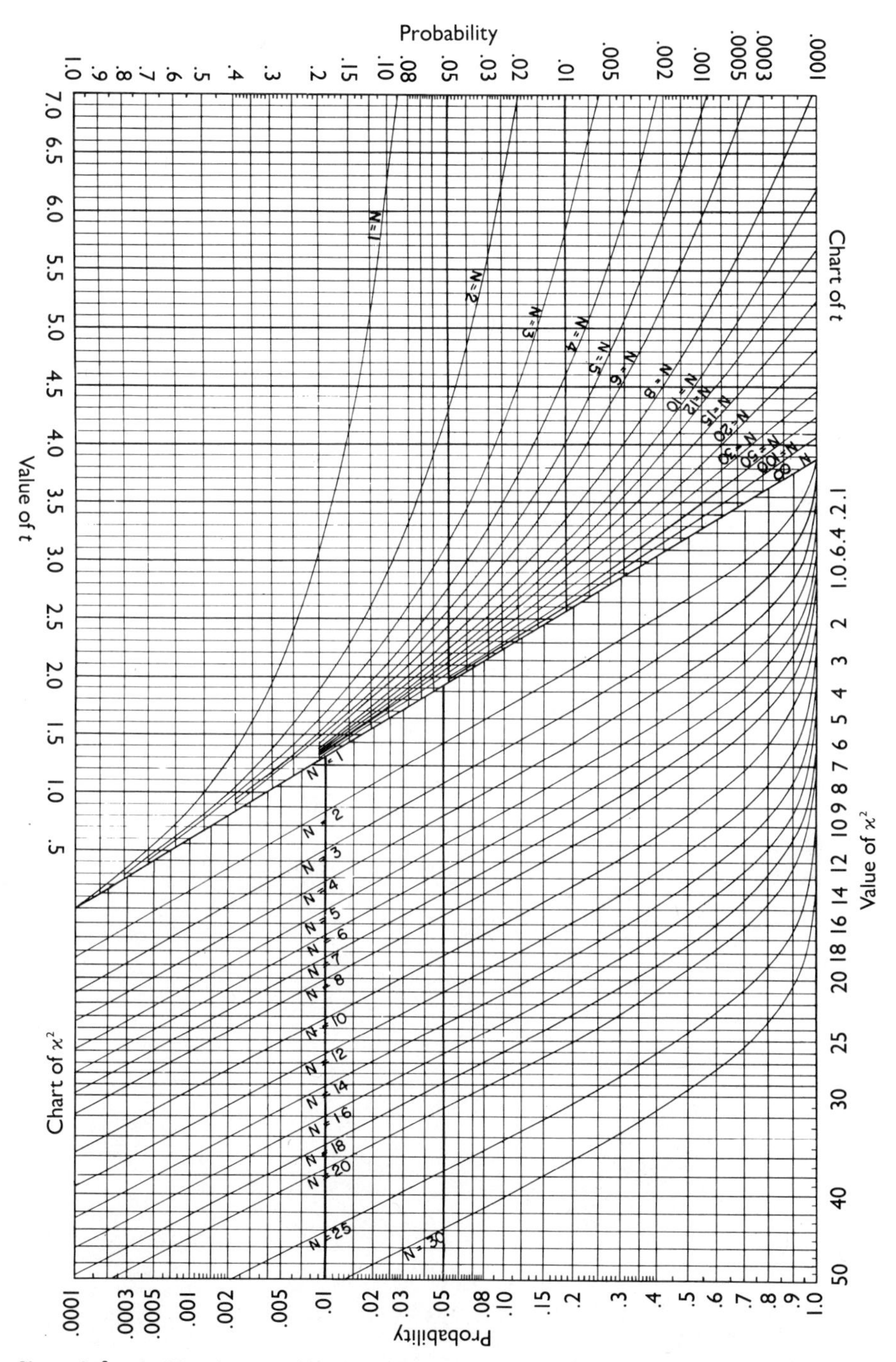

Chart of χ^2 and t ($N =$ degrees of freedom). To obtain the probability associated with a given value of χ^2, find the value along the bottom, go up to the diagonal corresponding to the number of degrees of freedom, and then read the probability at the left. To use the chart for a t test, turn it upside down, and follow the same procedure. (From J. F. Crow. 1945. *J. Amer. Stat. Assn.* 40:376.)

SELECTED REFERENCES

TEXTBOOKS ON POPULATION GENETICS

Cavalli-Sforza, L. L., and W. F. Bodmer. 1971. *The Genetics of Human Populations*. New York: W. H. Freeman and Company. The definitive work on human population genetics, replete with data and worked examples.

Crow, J. F., and M. Kimura. 1970. *An Introduction to Population Genetics Theory*. New York: Harper & Row. Reprinted 1978. Minneapolis: Burgess. This is often referred to as a source of additional material and is abbreviated C&K in the text.

Ewens, W. J. 1979. *Mathematical Population Genetics*. Berlin: Springer-Verlag. An advanced book requiring considerable mathematical background.

Hartl, D. L. 1980. *Principles of Population Genetics*. Sunderland, MA: Sinauer Associates. An excellent, clearly written textbook for advanced undergraduate and graduate students. I believe Hartl's book the best way to supplement the material given here with another book at about the same level.

Hartl, D. L. 1981. *A Primer of Population Genetics.* Sunderland, MA: Sinauer Associates. Essentially an abridgement of the reference above. Very clearly written and easy to read.

Hedrick, P. W. 1983. *Genetics of Populations.* New York: Van Nostrand Reinhold. A somewhat more detailed version of much of the material in this book.

Li, C. C. 1976. *First Course in Population Genetics.* Pacific Grove, CA: Boxwood Press. Elementary; easy to read; all the algebraic steps spelled out in detail. Li's sympathetic understanding of his readers' difficulties has been apparent since the first edition, published in China in 1948.

Malécot, G. 1969. *The Mathematics of Heredity.* New York: W. H. Freeman and Company. A translation of the 1948 French classic that set a new standard of mathematical rigor in population genetics. Malécot's work suffered from being published in French in an obscure journal, but he is now appreciated as one of the real innovators. Not easy reading.

Nagylaki, T. 1977. *Selection in One- and Two-locus Systems. Lecture Notes in Biomathematics.* Vol. 15. Berlin: Springer-Verlag. I believe this is the best way to get a taste of the precision and rigor that now characterize the field. Very precisely and economically written.

Nei, M. 1986. *Molecular Evolutionary Genetics.* New York: Columbia University Press (in press). This book is particularly useful to those who want to know more about the methodology, particularly statistical tests. There are abundant numerical examples.

Roughgarden, J. 1979. *Theory of Population Genetics and Evolutionary Ecology: An Introduction.* New York: Macmillan. As the name implies, the emphasis is on the interplay between population genetics and ecology.

Spiess, E. R. 1977. *Genes in Populations.* New York: Wiley. Another book in the Li tradition. No algebraic steps are omitted, and there are numerous examples and applications to real data.

TEXTBOOKS OF QUANTITATIVE GENETICS

Chapman, A. B. (ed.) 1985. *General and Quantitative Genetics.* New York: Elsevier. Numerous authors have contributed to this full account of genetics as applied to animal breeding.

Falconer, D. S. 1981. *Quantitative Genetics,* 2nd ed. New York: Longman. I believe this is the best introduction to the subject. It is written especially for animal breeders.

Kempthorne, O. 1957. *An Introduction to Genetic Statistics.* New York: Wiley. Reprinted. 1969. Ames, IA: Iowa State University Press. The emphasis is on statistical methods, especially analysis of variance and least squares.

Mather, K., and J. L. Jinks. 1982. *Biometrical Genetics,* 3rd ed. London: Chapman and Hall. The emphasis is on experiments with plants capable of self-fertilization and with inbred lines available. Especially good for its discussion of scaling transformations.

TEXTBOOKS ON AGE-STRUCTURED POPULATIONS

Charlesworth, B. 1980. *Evolution in Age-structured Populations.* Cambridge: Cambridge University Press. An advanced treatise, with a full treatment of Mendelian populations with overlapping generations and an age structure.

Keyfitz, N. 1968. *Introduction to the Mathematics of Population.* Reading, MA: Addison-Wesley. A standard text on human demographic methods, with many examples.

Lotka, A. J. 1956. *Elements of Mathematical Biology.* New York: Dover. This is a revision of a classic, first published in 1925. Although mainly of historical interest, it is still very much worth reading. Lotka was the first to show the stable age distribution property.

Pollard, J. H. 1973. *Mathematical Models for the Growth of Human Populations.* Cambridge: Cambridge University Press. Somewhat more mathematical than Keyfitz.

PROBABILITY AND STATISTICS

Conover, W. J. 1980. *Practical Nonparametric Statistics,* 2nd ed. New York: Wiley. Methods of statistical analysis when normality cannot be assumed.

Feller, W. 1968. *An Introduction to Probability Theory and Its Applications,* Vols. I and II. New York: Wiley. When the first edition appeared, it was a major new departure, but quickly became standard. It shows the rigor and power of modern probability theory.

Fisher, R. A. 1958. *Statistical Methods for Research Workers,* 13th ed. Edinburgh: Oliver and Boyd. The classic book that revolutionized the use of statistical methods in biology.

Mood, A. M., and F. A. Graybill. 1963. *Introduction to the Theory of Statistics,* 2nd ed. New York: McGraw-Hill. A standard textbook that has long been widely used.

Snedecor, G. W., and W. G. Cochran. 1967. *Statistical Methods,* 6th ed. Ames, IA: Iowa State University Press. A cookbook approach; very useful.

Tukey, J. W. 1977. *Exploratory Data Analysis.* Reading, MA: Addison-Wesley. Rough-and-ready statistical methods for making sense out of real data.

SOME INFLUENTIAL BOOKS, OLD AND NEW

Fisher, R. A. 1930, 1958. *The Genetical Theory of Natural Selection.* New York: Oxford University Press (Clarendon Press). Revised paperback edition. 1958. New York: Dover. A masterpiece, perhaps the most influential book on natural selection theory since Darwin. It is hard going, but well worth the considerable effort required to read it.

Haldane, J. B. S. 1932, 1966. *The Causes of Evolution.* New York: Harper. Reprinted in paperback. 1966. Ithaca, NY: Cornell University Press. Haldane's masterpiece. Although written half a century ago, it remains one of the best introductions to evolutionary principles.

Hutchinson, G. E. 1978. *An Introduction to Population Ecology.* New Haven, CT: Yale

University Press. Charming, informative, and scholarly; full of the kinds of insights you would expect from the dean of population ecologists.

Kimura, M. 1983. *The Neutral Theory of Molecular Evolution.* Cambridge: Cambridge University Press. The case for evolution by random drift by the person who has done the most to develop the idea.

Simpson, G. G. 1953. *The Major Features of Evolution.* New York: Columbia University Press. A book on the historical features of evolution by the leading paleontological evolutionary theorist.

Wright, S. 1968, 1969, 1977, 1978. *Evolution and the Genetics of Populations.* Chicago: University of Chicago Press. A four-volume set, summarizing the life's work of one of the founders of population genetics, as well as the work of many others.

BIOGRAPHY

There are full-length biographies of each of the four pioneers to whom this book is dedicated. They all had colorful, adventurous lives — a far cry from the ivory tower stereotype.

Box, J. 1978. *R. A. Fisher: The Life of a Scientist.* New York: Wiley. An insightful biography by his daughter.

Carlson, E. 1981. *Genes, Radiation, and Society: The Life and Work of H. J. Muller.* Ithaca, NY: Cornell University Press. An admiring biography by a Muller student. Muller's life was a series of disappointments and disillusionments, but, eventually his greatness was rewarded by a Nobel Prize.

Clark, R. W. 1969. *JBS: The Life and Work of J. B. S. Haldane.* New York: Coward-McCann. Haldane was a remarkable polymath and possibly the most colorful of all biologists. If you doubt that, read this book.

Provine, W. 1986. *Sewall Wright and Evolutionary Biology.* Chicago: University of Chicago Press. This is both scholarly and interesting, based on Wright's voluminous publications and correspondence, and on many hours of taped interviews.

BIBLIOGRAPHY

Felsenstein, J. 1981. *Bibliography of Theoretical Population Genetics.* Stroudsburg, PA: Dowden, Hutchinson, and Ross. No less than 7982 articles, books, dissertations, and such are listed by authors and in a KWIK index. It is a challenge to find something that is not included.

COLLECTIONS OF CLASSIC PAPERS

The following are collected reprints of some of the most important papers, tied together with explanatory essays by the editors. All are part of a series entitled "Benchmark Papers in Genetics," published by Van Nostrand Reinhold, New York.

Hill, W. G. 1984. *Quantitative Genetics.*

Li, W. H. 1977. *Stochastic Models in Population Genetics.*

Milkman, R. 1983. *Experimental Population Genetics.*

Weiss, K. M., and P. A. Ballonoff. 1975. *Demographic Genetics.*

MISCELLANEOUS BOOKS AND JOURNAL ARTICLES REFERRED TO IN THE TEXT

Aoki, K. I. 1981. "Algebra of Inclusive Fitness." *Evolution* 35:659–663.

Bell, G. 1982. *The Masterpiece of Nature: The Evolution and Genetics of Sexuality.* Berkeley: University of California Press.

Chung, C. S., O. W. Robison, and N. E. Morton. 1959. "A Note on Deaf Mutism." *Ann. Hum. Genet.* 23:357–366.

Cochran, W. G. 1954. "Some Methods for Strengthening the Common χ^2 Tests." *Biometrics* 10:417–451.

Crow, J. F. 1979. "Gene Frequency and Fitness Change in an Age-Structured Population." *Ann. Human Genet.* 42:335–370.

Crow, J. F., and K. I. Aoki. 1982, 1984. "Group Selection for a Polygenic Behavioral Trait." *Proc. Natl. Acad. Sci.* 79:2628–2631, 81:6073–6077.

Crow, J. F., and T. Nagylaki. 1974. "Continuous Selective Models." *Theor. Pop. Biol.* 5:257–283.

Dykhuizen, D. E., and D. L. Hartl. 1983. "Selection in Chemostats." *Microb. Rev.* 47:150–168.

Gould, S. J., and N. Eldredge. 1977. "Punctuated Equilibria: The Tempo and Mode of Evolution Reconsidered." *Paleobiology* 3:115–151.

Haldane, J. B. S. 1964. "A Defense of Beanbag Genetics." *Persp. Biol. & Med.* 7:343–359.

Hamilton, W. 1964. "The Genetical Evolution of Social Behaviour. I and II." *J. Theor. Biol.* 7:1–52.

Hardy, G. H. 1940, 1967. *A Mathematician's Apology.* Cambridge: Cambridge University Press. Hardy's opinionated, idiosyncratic, and thoughtful view of mathematics. The 1967 edition has a forward by C. P. Snow that you would also enjoy.

Kimura, M., and J. F. Crow. 1964. "The Number of Alleles That Can Be Maintained in a Finite Population." *Genetics* 49:725–738.

Kojima, K. 1959. "Stable Equilibria for the Optimum Model." *Proc. Natl. Acad. Sci.* 45:989–993.

Maynard Smith, J. 1978. *The Evolution of Sex.* Cambridge: Cambridge University Press.

Mourant, A. E., A. C. Kopéc, and K. Domaniewska-Sobczak. 1976. *The Distribution of the Human Blood Groups and Other Polymorphisms.* New York: Oxford University Press.

Nei, M. 1975. *Molecular Population Genetics and Evolution.* Amsterdam: North-Holland/American Elsevier.

Newman, H. H., F. N. Freeman, and K. J. Holzinger. 1937. *Twins: A Study of Heredity and Environment.* Chicago: University of Chicago Press.

Wright, S. 1922. "Coefficients of Inbreeding and Relationship." *Amer. Natur.* 56:330–338.

Wright, S. 1922. "The Effects of Inbreeding and Crossbreeding on Guinea Pigs." *Bull. U.S. Dept. Agric.* 1121:1–59.

Wright, S. 1951. "The Genetical Structure of Populations." *Ann. Eugen.* 15:323–354.

Wright, S. 1982. "The Shifting Balance Theory and Macroevolution. *Ann. Rev. Genet.* 16:1–19.

ANSWERS TO QUESTIONS AND PROBLEMS

CHAPTER 1

1. $2p_Mp_N = 10p_N^2$, $p_M = 5p_N$, $p_N = 1/6$.
2. The quadratic equation $2p_M(1 - p_M) = 0.42$ has two solutions: 0.3 and 0.7. Given that *M* is more common than *N*, the *M* frequency is 0.7.
3. (1) The gene-counting method does not require the Hardy-Weinberg assumption. (2) It uses all three data items instead of only two; the square root method does not use the frequencies of *NN* and *MN*, only their sum. (3) If you use the maximum likelihood method, it gives the gene-counting answer.
4. If the allele frequencies differ in the two sexes; it then takes two generations instead of one to arrive at H-W ratios. (In fact, with an X-linked locus and unequal starting frequencies in the two sexes, the approach to equilibrium is asymptotic. See C&K, pages 45 ff.)
5. That the parents of the present population mated randomly, but that this random mating had not been true at some earlier time. Perhaps the population is derived from immi-

grants from areas with different allele frequencies that have not been mating at random long enough to reach linkage equilibrium.

6. Half of the children of an *MN* mother are *MN*, regardless of the father's genotype. Since the proportion of *MN* genotypes is the same in both generations, half of the heterozygous children came from heterozygous mothers.

7. The proportion of all matings that are $MM \times MM$ is p_M^4, and all the children are *MM*. Since the total proportion of *MM* in the population is p_M^2 those from $MM \times MM$ matings are a fraction p_M^2 of the total.

8. The conditional probability that a person is heterozygous, given that he or she is of the dominant phenotype ($A-$), is $P(Aa|A-) = 2pq/(1-q^2)$, where q is the frequency of the recessive allele. In the $aa \times A-$ mating the probability of a recessive child is 1/2 if the dominant parent is heterozygous, so that altogether the probability is $(1/2)(2pq)/(1-q^2) = q/(1+q) = x$. In the $A- \times A-$ mating, if both parents are heterozygous, the probability of a recessive child is 1/4, so that altogether the probability is $(1/4)[2pq/(1-q^2)]^2 = [q/(1+q)]^2 = x^2$.

9. If p is the frequency of the gene for horns and $q = 1 - p$, the frequency of horned males, 0.96, is $p^2 + 2pq = 1 - q^2$; so $q^2 = 0.04$, $q = 0.2$, and $p = 0.8$. The proportion of horned females is p^2, or 0.64.

10. From Table 1-13, on hypothesis 1 the expected frequencies of $0 \times AB$ is equal to that of $A \times B$, which equals $p_a^2 p_b^2 (1 - p_a^2)(1 - p_b^2)$.

11. The allele frequencies are $p_A = 0.35 + 0.15 = 0.5 = p_B$.
(a) From Equation 1-5 the frequency of chromosome *AB* next generation is given by $P(AB) - 0.25 = (1 - 0.2)(0.35 - 0.25)$; $P(AB) = 0.33$.
(b) By the same method, $P(Ab) = 0.17$; $P(AB/ab) = 2(0.33)(0.17) = 0.112$.
(c) $P(AB) = P(ab) = 0.25$ at equilibrium; so the proportion of $AB/ab = 2(0.25)(0.25) = 0.125$.
(d) AB/ab and Ab/aB are each 0.125; so the double heterozygote frequency is the sum, or 0.25.

12. Let $P_{AB} = p_A p_B + D_{AB}$, $P_{Ab} = p_A p_b + D_{Ab}$, and so on. Then, noting that $P_{AB} + P_{Ab} = p_A$, and substituting from the first sentence into the left side leads, after simplification, to $D_{AB} = -D_{Ab}$. In a similar manner, $D_{AB} = -D_{aB} = -D_{Ab} = D_{ab} \equiv D$. Now this D is the same as the D in the problem, as can be seen by noting that $P_{AB}P_{ab} - P_{Ab}P_{aB} = (p_A p_B + D)(p_a p_b + D) - (p_A p_b - D)(p_a p_B - D)$. Expanding the right side and noting that $p_A + p_a = p_B + p_b = 1$, it simplifies to D. Hence the D in the problem is the same as D_{AB}, D_{ab}, $-D_{Ab}$, and $-D_{aB}$. Since, for example, $D_{AB,t} = (1 - r)D_{AB,t-1}$ (see Equation 1-5), D also decreases each generation by a fraction equal to the recombination rate.

13. (a) The frequency, p, of the F allele is $928/1158 = 0.80$. The (expected) heterozygosity is $2p(1-p) = 0.32$.
(b) For Gpdh and Amy, $B_{FF} = 816/1158 = 0.705$, $B_{SS} = 0.022$, $B_{FS} = 0.097$, $B_{SF} = 0.176$; $D = (0.705)(0.022) - (0.097)(0.176) = -0.0016$. For Gpdh and inversion, $P_{F-} = 0.723$, $P_{S+} = 0.028$, $P_{F+} = 0.079$, $P_{S-} = 0.171$; $D = 0.00674$. For Amy and inversion, $D = -0.012$. None of the pairs shows any appreciable linkage disequilibrium.

14. If the baldness locus is on the X chromosome, sons would inherit the allele from their mother and the father's genotype is irrelevant; so the expected proportion of bald fathers would simply be the population average, or 0.133. So the X-linked hypothesis does not fit. Y-linked inheritance is ruled out by finding *any* nonbald sons of bald fathers.

If the trait is recessive, let q be the frequency of the allele for baldness. Its frequency is the square root of 0.133, or 0.36. The frequency of bald fathers with bald sons is

$$P(\text{bald father})P(\text{bald son}|\text{bald father}) = q^2q = q^3.$$

Since the frequency of bald males is q^2, the proportion with bald fathers is $q^3/q^2 = q = 0.36$, which does not agree with the data.

Let us try dominant inheritance. Let p be the frequency of the allele for baldness; then $1 - p = q$ is the square root of $1 - 0.133$, or 0.931. It is easier to ask for the proportion of bald sons with nonbald fathers:

$$P(\text{nonbald father})P(\text{bald son}|\text{nonbald father}) = q^2p.$$

Since the frequency of bald males is $1 - q^2$, the proportion of bald males with nonbald fathers is $q^2p/(1 - q^2) = q^2/(1 + q) = 0.44$. Thus the proportion of bald males whose fathers were bald is $1 - 0.44$, or 0.56. The agreement is almost too good to be true. So these data argue for an autosomal dominant allele causing baldness. (The allele is known to be expressed only when there is a sufficient level of male hormones.)

15. $(0.02)(0.06) = 0.0012$.

16. The departure from linkage equilibrium each generation is $(1 - 0.033)$ times the value in the previous generation. The reason for 0.033 rather than 0.05 is that crossing over of X chromosomes happens only in females, and 2/3 of the X chromosomes are in females. (It would require about 21 generations, or some 600 years, to reduce the disequilibrium by half.)

17. (a) $[(q_1 + q_2)/2]^2$.
(b) $[(4q_1 + q_2)/5]^2$.

18. (a) $P(AB) = 0.35$; $P(Ab) = P(A) - P(AB) = 0.35$; $P(aB) = P(B) - P(AB) = 0.25$; $P(ab) = 1 - P(AB) - P(Ab) - P(aB) = 0.05$; $P(AB/AB) = 0.35^2 = 0.1225$; $P(AB/ab) = 2(0.35)(0.05) = 0.035$; $P(Ab/aB) = 2(0.35)(0.25) = 0.175$, and so forth.
(b) The fact that the two double heterozygotes are not equally frequent tells us that the population is not at linkage equilibrium, as does the fact that each haplotype frequency is not the product of its two allele frequencies.

From Problem 12, $D = (0.35)(0.05) - (0.35)(0.25) = -0.07$.

19. The F_1 are all double heterozygotes, $A\,B/a\,b$. In a randomly mating population only part of the population are of this classification. Since a crossover between the A and B loci makes no difference except in double heterozygotes, only a part of the crossovers are effective in reducing linkage disequilibrium in a randomly mating population, whereas all are effective in an F_1 population.

20. (a) $(x^2 + y^2)/2 = (x^2 + y^2 + xy - xy)/2 = [(x + y)/2]^2 + [(x - y)/2]^2$, which is larger than $[(x + y)/2]^2$, since it contains another positive term (unless $x = y$).
(b) If we let x and y be the recessive allele frequencies in two equal-sized populations, this inequality shows that the frequency of recessive homozygotes is always greater before amalgamation than after (unless the two populations have the same allele frequency). See Section 1-4.

21. The expected heterozygosities in the five populations in Table 1-2, from top to bottom, are 0.111, 0.370, 0.494, 0.390, 0.245, and the unweighted average is 0.322. The unweighted mean allele frequencies are $p_M = 0.4896$ and $p_N = 0.5104$, and the expected heterozygosity is 0.500. If the populations were pooled, the average heterozygosity would increase by about 0.18, or about 55 percent.

22. There are n homozygotes and $n(n - 1)/2$ heterozygotes, or a total of $n(n + 1)/2$.

23. Yes, provided the chromosome is long enough for the recombination fraction to be 1/2. (Of course, this situation would not be true for male *Drosophila*, female silk moths, or other places where recombination does not occur.)

24. (a) 1/4.
(b) Increase.

25. (a) $2BC + 2AD$.
(b) $2BC$.

CHAPTER 2

1. From the pedigree, F = 1/32. The recessive allele frequency $q = 1/100$. The risk is $q^2(1 - F) + qF = 0.0004$, a fourfold increase.

2. (a) $F_A = 1/4$. $F_I = 2(1/8) + 2(1/32)(5/4) = 21/64$.
(b) Again, $F_A = 1/4$. $F_I = 1/4 + (1/8)(5/4) = 13/32$.

3. If K is inbred, the pedigree as given is not complete. One procedure is to construct a fictitious ancestry of K that gives her an inbreeding coefficient of 1/4 (for example, by making her mother and father sibs). Then $F_I = 1/16 + 1/32 = 3/32$, compared with 1/16 if K were not inbred.

4. The number of lethal equivalents per gamete, since $F = 1/16$, is $(16)(0.05) = 0.8$; the number per zygote is then 1.6.

5. No, because with this many lethal equivalents the increased contribution of gametes from self-fertilization would not compensate for the deleterious effect of inbreeding. See Section 2-3.

6. Letting q be the frequency of the A_S allele, the frequency of homozygotes under the kin-mating system was $q^2(1 - F) + qF = 17/800$. Later, when $F = 0$, $q^2 = 0.01$. Thus $q = 0.1$, and substituting this value into the first equation gives $F = 1/8$.

7. F_{IJ} is 1/2 for brothers, 1/4 for brother and sister, and 3/8 for sisters.

8. $N/4$. It is reduced by half because only females contribute mitochondrial genes, and again by half because mitochondria are haploid. Note that this answer is also correct for a gene on the Y chromosome.

9. Yes, because of greater population mobility, mates are less likely to be related. Thus the average inbreeding coefficient is less, and therefore the proportion of homozygous recessives is decreased (assuming, reasonably, that the allele frequencies have not changed appreciably).

10. One, since they have identical Y chromosomes.

12. (a) The population may have been enumerated at an earlier stage, before mortality reduced the number of adults; only fertile adults contribute to the next generation, and infertile ones may have been counted; because of differential fertility the number of progeny per parent may be greater than random; the sex ratio may not be $1:1$.
(b) If each parent produces exactly the same number of progeny, the effective number will be greater than the actual number (in fact, approximately twice as great as when the number of progeny per parent has a binomial distribution).

13. The harmonic mean of these numbers is $1/[(1/10 + 1/50 + 1/250 + 1/1250)/4] = 32$, much closer to the smallest number than to the largest.

14. The kinship coefficient is the same as for an X-linked locus in diploids.
(a) For sisters, 3/8.
(b) For mother and daughter, 1/4.
(c) 1/2.

15. 3/8.

16. No, with strict neutrality the heterozygotes should increase with the population number (see Equation 2-13).

17. No, if the mutation rate is constant per generation rather than per year. (The points raised by this and Question 16 are discussed in Chapter 7.)

18. (a) Approximately $q/4$, where q is the frequency of the recessive allele.
(b) There are probably two or more loci, at each of which there is a recessive allele capable of causing the disease.

19. The value approaches 4 as the number of females increases.

20. Alkaptonuria is a much rarer disease than cystic fibrosis. Hence consanguineous matings produce a larger fraction of the incidence.

21. About 8.

22. $(1 - p_1^2 - p_2^2 - p_3^2 - p_4^2)(1 - F)$.

23. For the formula to measure absolute heterozygosity, each new mutant allele must be of a type not currently existing in the population.

24. (a) This occurs when an individual is homozygous for two independent alleles that happen to be in the same state.
(b) This situation could occur if one (or both) of the alleles had mutated during its descent from a common ancestral gene.

25. The heterozygosity is $(1 - 1/21)^5$, or 0.784, of the original amount; the reduction is

0.216. A cousin mating reduces heterozygosity by 1/16, or 0.0625. So the weight reduction would be 0.216/0.0625 = 3.46 pounds.

26. (a) From second cousins ($F = 1/64$) the estimated number of lethal equivalents per gamete is (64)(0.06 − 0.05) = 0.64. From 3/2 cousins ($F = 1/32$) the estimate is (32)(0.02) = 0.64. From first cousins ($F = 1/16$) the estimate is (16) (0.04) = 0.64.
(b) Since each of these calculations leads to the same estimate, they are consistent with the linearity of increase in death rate with inbreeding.

27. This mating is equivalent to self-fertilization. The heterozygous class would decrease by half each generation.

28. With inbreeding, the frequency of homozygous recessives increases from q^2 to an eventual value of q, which could be a very large increase. If the inbreeding is mild, the rate of increase is very slow, however. With assortative mating the value quickly approaches an equilibrium not very much larger than q^2. It does not go far, but it gets there in a hurry. See Table 2-2.

CHAPTER 3

1. If p is the allele frequency, $p = 0.9 + (0.97^{10})(0.1 - 0.9) = 0.31$, from Equation 3-3.

2. Start with Equation 3-5 and neglect terms that are much less than 1.

3. The population may not be in strict Hardy-Weinberg proprotions, particularly if the population is spread over a large area.

4. We already computed H_t and the average of H_S (Problem 21 in Chapter 1). Using these values, we get $G_{ST} = (0.500 - 0.322)/0.500 = 0.356$.

5. For H to be 1/2, $4N_e m$ must be 1, or $2Nm = 1$, which means that the absolute number of migrants per generation, Nm, is 1/2.

6. Here is a rough qualitative argument. The effect of a single migrant in increasing heterozygosity decreases in proportion to the population size; but the decrease in heterozygosity from inbreeding also decreases as the population size increases, so that the ability of a migrant to offset the decrease in heterozygosity is the same regardless of size. Perhaps you can think of a better argument.

8. For two alleles, from Equation 3-8, $H = 2N_e m/(1 + 4N_e m)$, which for $N_e m = 50$, 5, 0.5, and 0.05 gives $H = 100/201$, 10/21, 1/3, and 0.1/1.2.

CHAPTER 4

1. $s = 1/2$, $hs = 1/6$, $h = 1/3$.

2. (a) 2500 generations, since with slow selection the time required is inversely proportional to s.
(b) and (c) There is no such linear relationship; the answer would have to be worked out.

3. They are all equal.

4. The frequency, q, of the less fit allele is $h/(1 + 2h)$. Notice that s cancels out.

5. (a) Since $s = 1$, the required mutation rate, from Equation 4-20, is 0.0004, which seems high.
(b) From Problem 4, $q = 0.02 = 2h/(1 + 2h)$; $h = hs = 0.021$.

6. 1/196.

9. If the viabilities are in geometric ratio (for example, 1, x, and x^2 for *aa*, *Aa*, and *AA*), the H-W ratios are preserved.

10. $1 - st/(s + t)$.

11. (a) From Equation 4-27 the probability that her brother is a new mutant is 1/12.
(b) If you write $\mu_m = 10\,\mu_f$, you will see that the mutation rate cancels out. Only the ratio of the two rates is needed.

12. $p' = p/(1 - qs)$; $q = \mu/s$.

13. $p' = p/(1 - sq)$; $q = \mu/s$. Note that these equations are the same as the haploid formulae.

14. (a) If we assume that $0 < s \leq 1$ and if h is in the range 0 to 1, A' will be lost. If $h > 1, A'$ may or may not be lost depending on which side of a stable equilibrium the population starts from.
(b) $h > 1$.
(c) $h < 0$.

15. $D_1 = 0.0008$, $D_2 = 0.0002$, $D_3 = 0.0001$, $T = -0.05$. Therefore the equilibrium is stable. The equilibrium allele frequencies are 0.727, 0.181, and 0.091.

16. One way is to divide the figure into three triangles by drawing lines from the internal point to the three corners. Then each triangle has an area equal to half its base, which is a side of the original triangle, times its altitude. Since the three triangles have the same area as the large triangle does, whose area is half its base times altitude, the three altitudes must add up to the altitude of the large triangle.

17. (a) It is not consistent, since the mutation rate is not the square of the allele frequency.
(b) Since $q = \mu/h$, $h = 0.01$.

18. 0.

CHAPTER 5

1. The mean of the parents is 90 kg, 10 kg below the average. The progeny would deviate from the mean only 0.4 as much as the parents; so their expected weight is 96 kg.

2. (a) The narrow-sense heritability is $0.08/2F = 0.08/0.25 = 0.32$. The genic variance is $0.32(200) = 64$ bu^2.
(b) $200 - 80 = 120$ bu^2.
(c) 80 bu^2.
(d) 0.32.
(e) $120/200 = 0.6$.

3. (a) He is 10 cm above average, of which 8 cm are estimated to be genetic and 2 cm environmental. In an average environment he would be expected to be 8 inches above the average, or 178.
(b) The man's breeding value is 176, or 6 cm above the average. The woman's deviation is 0. The child would then be 3 cm above the average, or 173.
(c) $173 + 2 = 175$.

4. If the loci act additively, the $F_2 = (P_1 + 2F_1 + P_2)/4$. In this case $(10 + 40 + 100)/4 = 37.5$, which is not very close to 25. If we take the logs, we get $(1.00 + 2.60 + 2.00)/4 = 1.40$; $\log 25 = 1.40$. These results argue that the genes act multiplicatively between loci.

6. If the half-sibs were reared in independent environments this would imply a heritability greater than one, which is impossible. The results might occur as a statistical accident, of course. If the half-sibs were reared in the same home, this high correlation could be the result of a strong environmental influence of the home.

7. This could happen if there were a negative covariance between genotype and environment.

8. (a) Bilineal.
(b) Double half-cousins, for example.

9. (a) Maternal influences, e.g., common environment or cytoplasmic factors.
(b) Dominance.

10. From Equation 5-29 the equilibrium genic variance is $V_0/[1 - (0.79)(0.28)] = 1.28V_0$, where V_0 is the original genic variance.

11. Since the trait is completely genetically determined, the broad-sense heritability is surely 1. Since there is no correlation between relatives (except identical twins), the narrow-sense heritability is 0.

12. No, since to progress past the extreme of the original population with an asexual population, new mutations would be required. In a sexual population this can happen by recombination alone. How important mutations are in long-time selection is not known, but they would be the same in both sexual and asexual strains. (On the other hand, the rate of progress in the early generations would have been faster, since selection progress would be determined by broad- rather than narrow-sense heritability.

13. $(1/2)V_G + (1/4)V_D = (1/2)(0.6) + (1/4)(0.2) = 0.35$.

14. For broad-sense heritability both have a value 1 (assuming that the trait is completely genetically determined). For narrow sense the rate dominant is higher. As the allele frequency approaches 0, the heritability of the dominant trait approaches 1, whereas that of the recessive approaches 0. To demonstrate this, show that with complete dominance the narrow-sense heritability is $2q/(1 + q)$, where q is the frequency of the recessive allele.

15. The same as for a Mendelian population, except replace the narrow-sense heritability by the broad sense. (There is only one parent, so that the average of the parents is simply the value of this individual.)

16. (a) As explained in the chapter, in a two-allele system, selection changes the frequency of some alleles in the direction of 1/2, thus increasing their contribution to the variance, and others in the other direction, decreasing their contribution. These changes approximately cancel, at least in the early stages of selection.
(b) This would not be true in an asexual population, which soon runs out of variance.

17. Sib correlations include 1/2 the additive variance and 1/4 the dominance, and since the fractions are different, you cannot compute either narrow- or broad-sense heritability.

CHAPTER 6

1. (a) $n_{11} = 120(5/6) = 100$; $n_{21} = 50(3/5) = 30$; $n_{31} = 30(2/3) = 20$; $n_{01} = 100(3/5) + 30(2/3) + 20(1/2) = 90$; $N_1 = 240$.
(b) $R_0 = (5/6)(3/5) + (1/2)(2/3) + (1/3)(1/2) = 1$.
(c) $T = (1)(1/2) + 2(1/3) + 3(1/6) = 5/3$.
(d) $\lambda^3 - (5/6)(3/5)\lambda^2 - (1/2)(2/3)\lambda - (1/3)(1/2) = 0$; $\lambda^3 - (1/2)\lambda^2 - (1/3)\lambda - (1/6) = 0$.
Perhaps you will notice that $\lambda = 1$ satisfies this equation, or you might have guessed this from the fact that $R_0 = 1$. In any case since there is only one positive root, you can solve the equation by trial and error.
(e) Since $\lambda = 1$, n_0, n_1, n_2, and n_3 are in the ratio

$$1 : \frac{5}{6} : \frac{1}{2} : \frac{1}{3}.$$

(f) Since $\lambda = 1$, the answer is the same as that in (c).

2. (a) For both populations the equation for λ is $\lambda^2 - \lambda - 1 = 0$, with the positive root $\lambda = 1.6$.
(b) A: $n_1 = n_0(1/2)\lambda^{-1}$; $n_2 = n_0(1/3)\lambda^{-2}$.
B: $n_1 = n_0(1/3)\lambda^{-1}$; $n_2 = n_0(1/4)\lambda^{-2}$.
(c) $R_0 = 2$ for both populations.
The two populations have the same number of offspring per parent and the same rate of increase but differ in equilibrium age distribution.

3. (a) $E_0 = \int_0^\infty l(x)\,dx = \int_0^\infty e^{-cx}\,dx = 1/c = 50$ time units.

(b) $R_0 = \int_0^\infty l(x)b(x)\,dx = \int_0^\infty e^{-cx}k\,dx = k/c = 1.5$.

(c) $T = \frac{1}{R_0}\int_0^\infty l(x)b(x)\,dx = (c/k)\int_0^\infty xe^{-cx}k\,dx = 1/c = 50$.

(d) $\int_0^\infty e^{-cx}k\, e^{-mx}\, dx = 1;\ k \int_0^\infty e^{-(c+m)x}\, dx = k/(m+c) = 1;\ m = 0.01.$

(e) $n(x) = n(0)e^{-cx}e^{-mx} = n(0)e^{-(m+c)x} = n(0)e^{-(0.03)x}.$

(f) $\bar{x}_r = \int_0^\infty e^{-mx}l(x)b(x)\, dx = k \int_0^\infty e^{-(m+c)x}\, x\, dx = k/(m+c)^2 = 33.3.$

4. (a) Since the value of λ is 1, the population will neither increase nor decrease.
(b) It will not attain a stable age distribution but will oscillate.

5. The second.

6. (a) $(4 \times 10^9)(1.02)^{24} = 6.43 \times 10^9$.
(b) $(\log 2)/(\log 1.02) = 35$ years.

7. (a) From Equation 6-23, letting $N_t = K/2$ and $N_0 = 1$, we get $t = (1/r)\ln(K = 1)$.
(b) Taking the derivative of dN/dt with respect to N in Equation 6-22 and setting this value equal to 0, we get the solution, $N = K/2$. This is obviously a maximum rather than a minimum.

8. (a) In a growing population the average age of reproducing females will be less, because a growing population has more young persons.
(b) They will be the same when the intrinsic growth rate is 0.

9. This is reasonable if the requisite genes for such behavior exist. A more pedestrian explanation is that the very young and very old are easiest to kill.

11. Consider the size of the population at a time during the year 2026.

CHAPTER 7

1. The minimum number of changes is $49 + 2(29) = 107$; the proportion is $107/139 = 0.770$. As expected, this value is somewhat less than the estimated mean number, 0.824, since 0.77 is a minimum estimate.

2. All of these should be, and are, faster than nonsynonymous changes in the coding region. I would expect the rates of all of these to be approximately the same, but for pseudogenes to be faster if there are differences among the regions, since pseudogenes presumably have no constraints on them.

3. (a) The time should be shorter for a favored mutant.
(b) Contrary to intuition, a mildly deleterious mutation, if it is fixed at all, goes at the same rate as a favorable one with the same value of $|s|$.

4. It should spread through the population and eventually replace the other Y chromosomes.

5. $(1/4)^{10}$, or about one in a million. The point of the question is to emphasize that a few generations of selection in a Mendelian population can produce a phenotype that is vanishingly rare to begin with.

6. It should not have any effect, unless the infant death caused the next child to be born sooner, in which case it would lead to an increase in female births.

7. Since brothers share half their genes, this is a good gambler's risk.

8. There are two popular hypotheses. One asserts that species go through bottlenecks and that even large populations have been small at some time in the last few thousand generations. A second hypothesis, more likely in my opinion, is that the great majority of isozyme mutations are *very* slightly deleterious. In a large population these mutations would be selected against; the smaller the population, the more are effectively neutral. In a very large population, then, the 5 to 15 percent observed heterozygosity represents mutation-selection balance or possibly some form of balancing selection.

APPENDIX I

1. $7436/148{,}462 = 0.050$, by the definition of probability (Equation A-1).

2. $4196/148{,}462 + 7436/148{,}462 = 0.078$, by the addition rule (Equation A-2).

3. (a) q^5, by the multiplication rule.
(b) $p^5 + q^5$, by the multiplication and addition rules.
(c) p^3q^2.
(d) $(5!/3!\,2!)p^3q^2$.
(e) $P(4 \text{ girls}) + P(5 \text{ girls}) = (5!/4!\,1!)q^4p^1 + q^5$.

4. $[556!/(315!\,101!\,108!\,32!)](9/16)^{315}(3/16)^{209}(1/16)^{32}$.

5. (a) The conditional probability that the woman is heterozygous, given that she has a normal phenotype, is 2/3. If she is heterozygous, then since her husband is necessarily heterozygous, the probability that the first child will be affected is 1/4. Altogether, the probability of the first child being affected is $(2/3)(1/4) = 1/6$.
(b) The prior probability that the mother is heterozygous, from Part (a), is 2/3. If she is heterozygous, the probability that all three children will be normal is $(3/4)^3 = 27/64$. The probability that the mother is homozygous normal is 1/3; if she is homozygous normal, the probability of all her children being normal is 1. Thus the Bayesian probability that the mother is heterozygous, given that she has produced three normal children, is $(2/3)(27/64)/[(2/3)(27/64) + (1/3)(1)] = 27/59$. If she is heterozygous, the probability of an affected child is 1/4; so the probability that the fourth child will be affected is $(27/59)(1/4) = 27/236$.

6. $P(80 \text{ green}, 20 \text{ white}) + P(79 \text{ green}, 21 \text{ white}) + \ldots + P(70 \text{ green}, 30 \text{ white})$. (In case you are curious, the answer is about 3/4.)

7. There are 7! ways of placing the seven genes on seven different chromosomes. There are 7^7 ways of placing the seven genes without regard to the number on any particular chromosome. Thus by the definition of probability the answer is $7!/7^7$, or about 0.006.

8. $(4/7)(1/8)/[(4/7)(1/8) + (3/7)(1)] = 1/7$.

9. (a) This is the probability that both will be boys or both girls, or $1/4 + 1/4 = 1/2$.
(b) $(0.513)^2 + (1 - 0.513)^2 = 0.50034$—hardly any change.

10. Let A stand for the event that the twins are dizygotic and B for the event that they are the opposite sex. We are given that $P(B) = 1/3$. We also know that $P(B|A) = 1/2$. From Equation A-4, $P(B) = P(AB) = P(A)P(B|A)$. Rearranging, $P(A) = (1/3)/(1/2) = 2/3$. Hence the proportion of monozygotic twins is 1/3.

11. Let S stand for same sex, M for monozygotic, and D for dizygotic. Then by Bayes' theorem $P(M|S) = P(M)P(S|M)/[P(D)P(S|D) + P(M)P(S|M)] = (1/3)(1)/[(2/3)(1/2) + (1/3)(1)] = 1/2$.

12. This will not change the sex ratio, although it will change the distribution of the sex ratio in families of different sizes. Here is one way to solve the problem. The number of males in the family is, of course, 1. The probability of a family of size 1 = 1/2, of size 2 = 1/4, of size 3 = 1/8, and so on. Therefore the expected family size is $(1/2)1 + (1/4)2 + (1/8)3 + (1/16)4 + \dots$, which sums to 2. Thus the proportion of males is 1/2.

APPENDIX 3

1. The probability of no change is e^{-m}, where m is the mean number of changes per amino acid. Equating this value to $(141 - 68)/141 = 73/141$ and taking logs gives $m = -\ln(73/141) = 0.66$. The rate per year is $0.66/(8 \times 10^8) = 8.25 \times 10^{-10}$, or roughly one substitution per billion years.

2. If the traits are independent, the expected number of one-egg, concordant pairs is $(100 \times 60)/200$, and we can compute the other three by subtraction from the marginal numbers. $\chi^2 = 9.524$, $df = 1$, $P = 0.002$.

3. Using ( ) for expectations based on a 9:3:3:1 ratio and { } for expectations based on independence, we obtain the following table:

	Purple	Red	Total
Long	226 (235.69) {247.45}	97 (78.56) {75.55}	323 (314.25)
Short	95 (78.56) {73.55}	1 (26.19) {22.45}	96 (104.75)
Total	321 (314.251)	98 (104.75)	419

Testing 9:3:3:1: $\chi^2 = 32.4$, $df = 3$, $P < 0.0001$.
Testing 3:1, purple:red: $\chi^2 = 0.58$, $df = 1$, $P = 0.45$.
Testing 3:1, long:short: $\chi^2 = 0.97$, $df = 1$, $P = 0.32$.
Testing independence: $\chi^2 = 34.7$, $df = 1$, $P < 0.0001$.

We conclude that the segregation at individual loci and viability are not significantly different from normal, but that the loci are not independent.

4. Problem 4: The expected numbers are 312.75, 104.25, 104.25, and 34.74. $\chi^2 = 0.47$, $df = 3$, $P = 0.93$. Problem 6: $\chi^2 = 1.33$, $df = 1$, $P = 0.25$. (This is the probability of deviating from the expected by 5 or more, whereas the earlier problem asked for the probability of deviating by 5 or less.)

APPENDIX 4

1. If we use the same garden pea characters and if the mating is $P\,L/p\,l \times p\,l/p\,l$, the observed numbers and expected proportions are shown in the following table:

	Purple, long	Purple, short	Red, long	Red, short	Total
Observed number	A	B	C	D	T
Expected proportion	$(1-r)/2$	$r/2$	$r/2$	$(1-r)/2$	1

The probability of the observed results is

$$P = \left(\frac{N}{A!\,B!\,C!\,D!}\right)\left(\frac{1-r}{2}\right)^{A+D}\left(\frac{r}{2}\right)^{B+C}$$

$$L = \ln P = \text{constant} + (A+D)\ln(1-r) + (B+C)\ln r.$$

Differentiating with respect to r, we get

$$\frac{dL}{dr} = -\frac{A+D}{1-r} + \frac{B+C}{r} = 0.$$

Solving this equation gives $r = (A+B)/T$, the customary formula used in linkage studies.

2. If the recombination amounts in the two sexes are r and r', then x in the formulae at the beginning of this section is rr'. Taking the square root gives the geometric mean of the recombination proportion in the two sexes rather than the arithmetic mean.

3. The observed and expected numbers are shown in the following table:

	Black	Brown	White	Total
Observed number	81	19	90	190
Expected proportion	$(1-r)/2$	$r/2$	$(1-r)/2 + r/2 = 1/2$	1

$$L = \ln P = \text{constant} + 81\ln(1-r) + 19\ln r$$

$$\frac{dL}{dr} = -\frac{81}{1-r} + \frac{19}{r} = 0 \qquad r = \frac{19}{100}.$$

Note that the answer agrees with common sense; since the white mice are uninformative, the maximum likelihood procedure ignores them.

4. From Table 1-13, if we let $x = p_a^2$ and $y = p_b^2$, the probability of the observed results is

$$P = (\text{constant})\{(xy)^{88}[(1-x)y]^{44}[x(1-y)]^{27}\,[(1-x)(1-y)]^{4}\}$$
$$= (\text{constant})\{x^{115}y^{132}(1-x)^{48}(1-y)^{31}\}$$
$$L = \ln P = \text{constant} + 115\ln x + 132\ln y + 48\ln(1-x) + 31\ln(1-y).$$

Taking partial derivatives, we get

$$\frac{\partial L}{\partial x} = \frac{115}{x} - \frac{48}{1-x} = 0 \qquad x = p_a^2 = \frac{115}{163}$$

$$\frac{\partial L}{\partial y} = \frac{132}{y} - \frac{31}{1-y} = 0 \qquad y = p_b^2 = \frac{132}{163}.$$

Thus the solutions given in Chapter I are maximum likelihood solutions.

5. If we let q be the expected proportion of males and q^2 that of females, the probability of the observed results (omitting the constant and simplifying) is

$$P = q^{725}(1-q)^{8324}q^{80}(1-q)^{9032}(1+q)^{9032}$$

$$L = 805 \ln q + 17{,}356 \ln(1-q) + 9032 \ln(1+q).$$

Equating the derivative of L, to 0 with respect to q gives

$$\frac{805}{q} - \frac{17{,}356}{1-q} + \frac{9032}{1+q} = 0$$

and the solution is $q = 0.0772$. You may want to verify that using expected numbers based on this estimate will give you a χ^2 of 4.77, which with one degree of freedom gives a probability of 0.03.

INDEX

MATHEMATICAL SYMBOLS

References to symbols give the page on which the symbol is defined or first used. Greek letters are alphabetized by their English spelling. Lower case symbols are given first.

AUTHORS

SUBJECT